S0-EJV-158

Infectious Cancers of Animals and Man

This book is dedicated to the memory of my father, the late Gerard Yorke Twisleton-Wykeham-Fiennes and of my mother Gwendoline. Gerard Fiennes, as a leading journalist was, in the early years of the century, keenly interested in the cancer problem and used his influence to raise money for cancer research with considerable success

Infectious Cancers of Animals and Man

RICHARD N. T-W-FIENNES

ACADEMIC PRESS 1982
A Subsidiary of Harcourt Brace Jovanovich, Publishers
London New York
Paris San Diego San Francisco São Paulo
Sydney Tokyo Toronto

ACADEMIC PRESS INC. (LONDON) LTD
24/28 Oval Road
London NW1 7DX

United States Edition published by
ACADEMIC PRESS INC.
111 Fifth Avenue
New York, New York 10003

British Library Cataloguing in Publication Data

T-W-Fiennes, Richard N.
Infectious cancers of animals and man.
1. Oncogenic viruses 2. Viral carcinogenesis
I. Title
599.02 QR372.06

ISBN 0-12-256040-X
LCCCN 82-71404

TYPESET BY OXFORD VERBATIM LTD, FARINGDON
PRINTED IN GREAT BRITAIN BY
T. J. PRESS (PADSTOW) LTD, CORNWALL

Preface

When one proposes a theme for a book to a publisher, one is asked to answer three questions: 1. Why is a book on this subject needed? 2. Who will buy and read this book, when it is published? 3. Why do you think you are the right person to write such a book? It is a wise precaution on the part of the author to forestall reviewers by attempting to answer these questions in a preface. To discredit him on these points is an obvious and all too frequent gambit on the part of hostile reviewers; friendly reviewers will genuinely wish to know. A short book on the infectious causes of animal cancers is an obvious need. No such book exists, the subject is little known, and it is not easy to find the basic information without studying massive tomes or reading papers in scientific journals, not always easily understood and legion in number. My proposed book was, therefore, to be short and simply written as an introduction to the subject; it does, however, include ample bibliographies which can lead the reader to more extensive studies. The book is not a comprehensive work on the infectious animal cancers in a textbook sense. It highlights those cancers, on which extended studies have rendered the infectious cause beyond contention.

It was difficult to decide how much to include about those cancers of man, which are thought to be or might be of infectious origin, since this subject is highly contentious. However, there are some human cancers, which are believed to be of viral origin and it would appear to be irrational to exclude them from this account. Even though the tentative conclusions reached may be disputed, the case should be presented together with the evidence relating to it. There is, furthermore, indisputable evidence that, as with other animals human genetic and somatic cells carry genes of some viruses, which are capable of causing cancer in other animals and cause oncogenic changes in human cells grown in culture.

The material here outlined is likely to be better known to my veterinary colleagues than to members of the medical profession, since the infectious causes of some cancers, particularly in poultry, have long been known and accepted; more recently the infectious causes of some cancers of cats and cattle have been adequately demonstrated and are fully accepted; more recently still, cancer problems in primates other than man have become of crucial importance. So long as the causes of human cancers remain in doubt, it would seem that information should be readily available in easily assimilable form about similar cancers in animals, since this could have relevance in the human field.

I would hope, therefore, that the material included in this book would prove of interest to medical students and graduates, and to all who see a challenge in this most mysterious group of diseases, including general biologists, geneticists and virologists. It is possible even that some persons without specialist scientific training could find the material of interest, if their reading was sufficiently selective; for this purpose, I have kept the presentation as non-technical as possible.

The cancers are today of such widespread importance and involve so much anguish for patients and their relatives and friends, that many people with broad interests and a comprehensive education might well be tempted to learn something of this aspect of the subject and of the formidable problems which delay a solution.

As to whether one is the right person to write such a book is a question which a prospective author, endowed with becoming modesty, may find more difficult to answer. It is a task, which, in spite of interest and dedication, I would have rightly hesitated to undertake until the supposedly leisured days of retirement. Except in sober retreat, the mass of literature on the subject could not be properly studied and analysed. To qualify to write this book, one would need to be: 1. A veterinary pathologist with specialist knowledge of animal cancers; 2. A human medical pathologist with specialist knowledge of human cancers; 3. A geneticist; 4. A virologist; 5. An immunologist; 6. A biochemist with special knowledge of cellular metabolism, and 7. A physicist with knowledge of X-ray crystallography. All these disciplines are involved in the cancer problem, and no one man could bridge them. Since nobody can write authoritatively on all these problems, I may be excused for doing the best I can with the skills I have acquired and developed. These include a Cambridge degree in the Natural Sciences (Botany, Zoology and Physiology) and an Edinburgh degree in Veterinary Medicine. As a zoo pathologist for nearly twenty years, specialising in comparative – and especially monkey – medicine, I have found myself with a foot in both the human and veterinary medical camps with probably a wider overview of overlapping problems than is given to most. One would suppose, therefore, that, if the task I have set myself should lie within the capacity of one person to perform, I should be able to perform it. However, *Capax imperii, nisi imperasset*!, which may be translated 'The proof of the pudding is in the eating!'

Note

Throughout this work, the nomenclature used for viruses is that pertaining when the material was being assembled, and that used in most of the works to which reference is made.

A revised nomenclature for viruses, as now internationally accepted, is given by Matthews, R.E.F. (1979). 'Classification and Nomenclature of Viruses,' Basel: Karger. In the new system, the term 'oncornaviruses' is replaced by 'oncoviruses', which become a subsection of the retrovirus group. This change applies to oncornavirus groups A, B and C. Oncornavirus group D, as with the Mason-Pfizer Monkey Virus (MPMV), becomes retrovirus D.

It was decided not to revise the nomenclature throughout this work, since to do so might have introduced more errors than it cured and the change would be confusing to those consulting works of reference used.

Contents

1
Prologue

In the United States, one insurable category of illness is that known as the 'dreaded diseases'; they are the cancers. So much are they dreaded that many, who believe themselves affected, are too scared to consult the doctor in case their fears are confirmed. There can be few people in Western societies, who have not at first or second hand been concerned with one or more of the tragedies that cancer imposes. It may be a child of eight years or more, a bright happy child, whose health and spirits begin to wane; she has developed leukaemia, and the parents are told that she has not more than a year to live. It may be a young mother with a lump in her breast; the breast is surgically removed together with the regional lymph glands; she is, perhaps, subjected to drug therapy which causes her hair to fall and she must wear a wig; perhaps she is lucky and survives; perhaps after an agony of uncertainty for herself and her family, she dies when still in her thirties. Older people, in whom the dreaded symptoms appear, may undergo repeated surgery and treatment with drugs and rays, until their pain is so great that they live the last months of their lives in a cloud of morphia and heroin.

There has been great progress in the treatment of the cancers, especially when it can be started early enough. In some leukaemia cases, a permanent cure rate is high and sufferers from other forms of cancer, too, have been permanently cured, or given a usefully prolonged lifespan. Moreover, means have been discovered whereby cases can be diagnosed at an earlier stage, when the outlook for treatment is more favourable. There is, therefore, no longer good cause for fear to consult the doctor, when doubts exist. Many persons, who would formerly have died, have been saved by early treatment; yet, too many cases succumb in spite of treatment, and often the treatment itself involves much pain and misery. In Western countries today, cancer comes second in the death league after episodes of the heart and arteries, which cause strokes and coronary thromboses. It is much more feared and, whereas cardiovascular diseases are probably consequences of certain ways of life, there are reasons to hope that the cancers may be overcome.

While the prospects for cancer patients are greatly improved, the ultimate causes of cancer in human beings are still unknown. It is said that cancers are caused by certain 'carcinogenic' chemicals, or by X-rays and other forms of radiation, by excessive smoking and other means. Indeed, there has been something of a witch hunt for chemicals that might cause cancer; even products such as saccharin, which have been in use for many years, have been incriminated. Undoubtedly, irradiation and the misuse of chemicals and tobacco are *associated* with the incidence of cancers; this is not to say that they *cause* them, and there are reasons for supposing that they may not. If they are avoided there is a better chance of avoiding cancer, but many heavy smokers do not develop cancer of the lung (bronchial carcinoma) and many who do not smoke become affected with this cancer.

Cancer is not a prerogative of the human race. Malignant growths occur in invertebrates, such as marine polyps; they occur in fish and frogs; birds are severely affected, and they have been studied in many species of mammals; the root cause of cancer in those animals, in which they have been studied, is believed to be known; in man it is not. In fishes, frogs, birds, mice, rats, hamsters, guinea pigs, rabbits, dogs, cats, cattle, monkeys and other animals, cancer is known to be the result of infection with certain viruses, against which in some cases effective vaccines can be prepared. There are plant cancers, too, and these are also the product of infectious causes. It might be supposed absurd to think that man's own cancers are different from all others in the plant and animal kingdoms, that his alone are caused by some factors, which are not transmissible; it is a pre-Darwinian concept, which places man apart from all other living species. However, a great many people do think this and, for reasons which will be discussed, scientific proof of the infectious origins of human cancers has been extraordinarily elusive. Yet, the means to control the incidence of cancer could depend on the production of this proof and on a knowledge of the infectious agents concerned and the means to manipulate them in such a way as to produce prophylactic and therapeutic agents from them.

Curiously, no simple account of the researches on the infectious causes of cancer has ever been written. There are a number of lengthy and scholarly works, such as those of Ludwik Gross (1970) and Edouard Kurstak and Karl Maramorosch (1974), which are too detailed for the uninformed reader and which require to be updated. It is my hope that this work will present the facts about the animal cancers in such a way that the reader can understand them, and draw his own conclusions about their relevance to the human situation.

In this chapter, therefore, I shall try to answer two questions: 1. What are cancers? and 2. What are viruses? The history of the cancers as infectious diseases will then be pursued in following chapters. It will no doubt come as a surprise to many that the first proof of an infectious cause of a cancer, in domestic fowls, was produced more than seventy years ago in 1908 in Copenhagen. It may also not be realised that an anti-cancer vaccine against a fowl tumour is in commercial production and used with great success.

Tumours commonly occur in all groups of animals; they are classified as benign and malignant. Benign tumours are usually growths of fibrous tissue; they are hard and have well defined borders. After a time, they cease to become further enlarged and remain as a lump that is unsightly but does no harm. Sometimes, they regress and disappear; very occasionally they become secondarily malignant. A number of such growths are known to be associated with virus infections. For example, myxomatosis in its natural host, the cottontail rabbit, causes transient swellings on the skin, though in European rabbits the same virus causes a rapidly fatal septicaemic disease. Warts in human beings are caused by a virus, the human wart virus. Malignant tumours on the other hand have diffuse borders and grow progressively at the expense of the tissue, in which they are situated. They are the true cancers or neoplasms, spreading out like a crab, destroying and replacing the tissue. They grow to such an extent that the blood supply at the centre becomes exhausted, and the cells become necrotic and die. Sometimes, especially in the early stages, the tumours remain localised; they can then be removed surgically with a reasonable hope of success and the residual cancer cells can be destroyed by radium or other suitable treatment. All too often, some malignant cells have been carried away in the blood or lymphatic vessels to other organs, where they start secondary tumours known as metastases. The first sites, where metastases are found, are usually in the lymphatic glands associated with the host tissue; the glands are, therefore, usually removed when the tumour is excised. Once metastatic spread has occurred, a great many vital organs may be affected; multiple tumours can grow throughout the body and the case becomes hopeless. Very occasionally, for reasons unknown, tumours regarded as hopeless disappear of their own accord and the patient recovers. Such remarkable occurrences have been the source of many claims for the efficacy of quack cures, which have over the years seriously embarrassed the medical profession. That such can happen gives obvious hope that there do exist some unknown factors which, if discovered, could be employed to effect cures. Newer knowledge of cancer viruses may suggest the reasons why this happens, and will be discussed later.

In classical pathology, diseases were classified in two groups: 1. inflammatory, and 2. non-inflammatory. The characteristics of inflammation were described as: 'rubor, dolor, tumor, turgor', that is redness, pain, swelling and turgidity. Tumours obviously exhibit swelling, but not redness, pain or turgidity; there is, of course, pain associated with cancer, but the swellings themselves are not painful. All infectious diseases are inflammatory, and so is injury, such as wounds, fractures and bruises. Cancers were regarded as non-inflammatory and non-traumatic, therefore non-infectious. To this day, many people find it difficult to associate the non-inflammatory processes of the cancers with infectious causes. This should be less difficult for those trained in veterinary medicine, who should know of the infectious tumours of birds, discovered so long ago, and more recently in cats and mice and other animals.

The major effort in cancer research was directed formerly to studying the nature and properties of the cancer cells, and to determine why they possessed their unusual properties of continuous growth. The multiplication of normal cells in a tissue is self-regulatory, so that they combine to form an ordered tissue within an organ such as the liver, for example. Cancer cells are 'rogue' cells and do not obey the rules. They are, however, still cells derived from the body's own tissues, which in technical parlance have been 'transformed'; though they have acquired special properties, the parent cells from which they have been derived can still be recognised. Cancers derived from 'transformed' blood cells are known as 'leukaemias', of which there is a wide variety; those derived from lymphatic cells are known as 'lymphomas'. Cancers derived from connective tissues are known as 'sarcomas'; from epithelial or endothelial lining cells as 'carcinomas'. There are therefore, three main groups of cancers: 1. of the blood, 2. of the connective tissue, and 3. of the lining tissues. Furthermore there are a great many other terms to describe tumours, such as 'osteomas' from bone, 'chondromas' from cartilage, 'adenomas' from gland cells, 'fibromas' from fibrous tissue, and so on. Sometimes, too, there are mixed tumours, such as 'fibro-sarcomas' or 'adeno-carcinomas'.

The identification of tumours is of great importance. The pathologist must first of all give his opinion as to whether tumour tissue is benign or malignant; if he can say with certainty, which he can usually do, it is no less important that its nature should be ascertained. Different tumours behave differently in the tissues they affect, in the speed at which they progress, and in their tendency to metastasise. It is important also in relation to the nature of the viruses which may be involved. Some viruses affect certain groups of tissues and certain base cells; others have different orientations. Some viruses are very restricted in the cells they can attack; others have a greater range, being known as 'pluripotent' or 'totipotent'. These matters and many complexities arising from them will be discussed in later chapters. Meanwhile, let us briefly discuss what viruses are and the difficulties involved in their study.

The first discovery of a virus was made by a Russian botanist, Ivanowsky, in 1892, when he passed the sap from tobacco plants affected with mosaic disease through an earthenware filter and infected healthy plants by rubbing the sap over their leaves. Since all cells and bacteria had been removed by the filter, evidently there was present in the filtrate an infectious agent too small to be detected by the light microscope. Such agents came to be known as ultra-microscopic or filterable viruses. Pasteur, himself, at an earlier date had suggested that such an agent was likely to be the cause of rabies, but had not demonstrated it. The first animal disease demonstrated to be due to a virus was foot and mouth disease in 1898, and the first human disease, yellow fever, was shown to be due to a filterable virus in 1901. Today virologists are equipped with a great many sophisticated techniques for the study of viruses. They can be seen under the electron microscope, so that their structure can be studied; they can be grown in tissue cultures; they can be precipitated by the ultra-centrifuge; and their composition can be determined by advanced chemical techniques. Even the genes they carry can be isolated and related to their function. At the time of the earlier studies of viruses, the electron microscope had not been invented, and the use of tissue cultures had to await the discovery of penicillin for the control of contaminants. The material under study was passed through various forms of porcelain or asbestos filters under sterile conditions, and was then tested in animals for infectivity. In the course of time, means were found to manipulate some viruses; live or killed vaccines were prepared, and immune sera were made available. In 1908, Ellerman and Bang in Copenhagen showed that diseases of the leukosis complex (the leukaemias) in chickens were the result of viral infections; in 1911 an American veterinary surgeon, Peyton Rous, transmitted solid tumours, sarcomas, of fowls by means of filtered material. As long ago, therefore as 1911, two groups of tumours of fowls were known to be caused by filterable viruses. The story proved to be complicated, and will be pursued in greater detail in ensuing chapters.

Viruses vary greatly in size, although they are so small; some, such as those of the poxvirus group, are comparatively large and little below the range of the light microscope; others, including some involved in the cancer story, are very small particles indeed. Viruses can only live and multiply inside living cells, on which they are totally dependent. As life forms they are incomplete in that they possess only one of the essential components of life, either desoxyribonucleic acid (DNA) or ribonucleic acid (RNA), but never both. Their DNA or RNA is packed in a limited number of genes, eight in the myxoviruses of influenza, which can mediate the synthesis of some protein enzymes; their chief function, however, is to manipulate the genetic material of the host cell so that it synthesises the materials necessary for the production of new virus particles. Some viruses are 'defective' even so, and cannot manipulate the host cell, unless a 'helper' virus is also present from which it can borrow additional genetic information. This is the case with one of the 'oncogenic' (cancer producing) viruses and was one of the stumbling blocks to their study, until this was realised.

Viruses are composed of an external coat of various bizarre shapes and patterns, which is constructed of protein in association with lipid (fatty) substances or other materials and is covered by a limiting membrane; within these structures are the genes. With RNA viruses, the genes within the limiting membrane become attached to the cell's limiting membrane or to the internal network of fibres ('endoplasmic reticulum'); in these situations more viral particles are produced by budding, the host cell providing both the energy sources and the material. The DNA viruses enter the nucleus of the host cell and become incorporated in the genetic material, where new virus particles are monitored. We shall be much concerned with one group of RNA viruses, the oncornaviruses, which produce an enzyme with the property of creating a DNA copy of the virus; by this means, the virus can enter the cell nucleus and become incorporated in the host genetic material. In this situation, these viruses are sometimes replicated as if they were DNA belonging to the host, unlike ordinary viruses; they divide, without harming the host cell, as if they were in fact host DNA, and remain incorporated in the daughter cells resulting from the division. They are frequently present in the germ cells and so can be passed from one generation to the next in the sex cells of either parent; indeed, there is evidence that some of these viruses have been passed from generation to generation for millions of years, evolving *pari passu* with the host, just as if they were host genes. Some workers, indeed, have suggested that they may be an important factor in evolution and that their activities are not necessarily inimical. In spite of this intimate relationship, they retain the power under certain conditions to revert to a replicatory phase, by which new viral particles can be created in the cell cytoplasm (the sap). They are also associated with a number of serious cancers. Detection of the presence of these latent viruses is attended with great difficulty, because they are not present in the form of virus particles only as viral genetic information; such information has been detected in cells of all animals studied, including man, at all ages and even in eight cell foetuses. Within the cell nucleus, the viral DNA may also indulge in gene exchange with the host, so that viral particles may emerge in which there are incorporated host genes; such may make them acceptable to further host cells and facilitate their entry into new cells that are not infected. Further, if the host cell should become additionally infected with another related virus, the two viruses may exchange genes, so that two hybrid viruses with new properties may be produced. This behaviour on the part of these viruses has made their study one of bewildering complexity.

In the normal course of events, replicated viral particles emerge from the host cell destroying it in the process. This process continues, until so many cells are destroyed that the organ to which they belong can no longer perform its function and the host dies. Alternatively, the host develops immune defences, by which the virus is killed and recovery takes place. Sometimes, however, the virus becomes inactive within the host cell as with the oncornaviruses. For example the virus of chicken pox, one of the herpesvirus group, acquired in youth becomes latent in certain cells of the nervous system and is harmless, but may re-emerge in later years to cause shingles; if a young person acquires the virus from a shingles case, he develops chicken pox. As a third alternative, some viruses which possess 'transforming' genes may cause neither acute symptoms of infection nor remain latent; they then cause the host cells to mutate and become cancerous. Such is mostly the result of aberrant behaviour and more likely to occur in other than the natural hosts of the virus. The mutagenic powers of the virus may, furthermore, require to be activated by 'co-carcinogens' such as carcinogenic chemicals or harmful rays. In addition, because of gene exchange, viruses lacking 'mutagenic' properties may acquire them from newly infecting related viruses, or newly infecting viruses may acquire mutagenic properties from other viruses, which are lying latent and do no harm. The gene responsible for cell transformation is distinct from that concerned with replication, so that the two activities do not go hand in hand.

Proponents of the viral causation of cancer consider that the behaviour of cancer cells is so wayward as to be explained only by an input into them of new genetic information; such an input, they reason, could only come from the incorporation of viral genes into the genome. Most, if not all, tissue cells studied from whatever animal, man included, can be shown to harbour either viruses or viral genes in masked form; these belong to the group of oncornaviruses, which are the most important in cancer causation. The viral theory is further supported, because cancers induced in fowls or mice by oncogenic chemicals or irradiation invariably shed oncogenic viruses when grown in tissue cultures, and the same viruses cause cell transformation, when transferred to new culture media or in healthy animals infected with them. The chemicals or irradiation, it is argued, are thus acting as co-carcinogens by impairing the powers of the cells to control their own resident viruses, which are then freed to cause cancer. In this way a latent virus inherited 'vertically' from one or both parents could cause cancer, even in late life. The opponents of these ideas admit that viruses can in some cases cause cancer, and they accept that in some cancers the circumstantial evidence is strong that they do so in man. They believe, however, that in most cancers it is not the viruses, which actually cause the cancer, but the damage that they do to the genetic control of the host. It is, they reckon, a derangement of the gene control of the host that is to blame, and that such derangement can equally well be caused by subjecting the cells to deleterious influences, such as oncogenic chemicals or harmful rays, the intervention of viruses being unnecessary.

If human cancers, or some of them, are indeed caused by viral infections, there would seem to be a good hope that means could be found to control those, at any rate, that are of the greatest importance; the argument is, therefore, by no means academic. This could well come about in an unexpected way, as is illustrated by the case of Marek's Disease in chickens, the only neoplastic condition at present routinely controlled by means of a vaccine. Marek's Disease affects young chicks and causes lymphatic cell tumours in the nervous system (neurolymphomatosis). It is very prevalent in some flocks of poultry and causes severe losses amongst the younger birds. It is caused by one of the herpesviruses, of which there exists a related form in turkeys. Protection is afforded by injecting the young chicks with the turkey strain of the virus; as a result the chicks became infected with the turkey virus, which they retain in latent form throughout their lives. However, they are not prevented by this from contracting infection with the poultry strain; on the contrary, they contract infection with the poultry strain and retain this virus also in their bodies throughout their lives. The poultry virus is, nevertheless, inhibited from causing malignant symptoms. How or why this happens is unknown, but it is unreasonable to suppose that similar phenomena do not occur also in other animals and with other cancers. Could it be that in those rare cases, when malignant tumours in human patients regress, some related but antagonistic virus has been fortuitously acquired by the patient? It is as good an explanation as any other. Possibly, hybrid viruses with altered properties are produced; possibly, it is simply a case of enhanced host control. Many questions are posed, requiring a greater knowledge of the viruses involved, their reactions with each other and with the tissue cells of the host.

Viral genes present an intriguing problem. They are limited in number and readily dissociate from each other within the host cell. Genes of related viruses are readily exchanged to form 'recombinant' viruses with hybrid properties. They even become exchanged with host genes, so that daughter viral particles may contain some host genetic information. They regularly and readily displace certain genes of the host and become implanted in the host genome, where they reside and behave as if they belonged to the host for thousands of generations. The 'incomplete' viruses even lack the gene for particle replication, and must borrow this gene from a 'helper' virus before multiplication can occur. The gene for cell transformation is of especial interest, but too little is known of its nature or of the conditions under which it remains inoperative, is activated, or manipulates the behaviour of the host cell. The possession of such genes is a rarity in the viral world. Amongst the many recognised groups of viruses, only four have been associated with different forms of cancers. Amongst these, oncogenic properties are only present in a few members of each group and they only cause cancers in some animals, not in others. Furthermore, viral groups possessing 'oncogenes' are widely dissimilar and not closely related to each other. The four groups responsible for causing cancer are: 1. the papova group, 2. the adenovirus group, 3. the herpes group, and 4. the oncornavirus group.

The papova group

This group consists of the papilloma viruses, mouse parotid tumour virus (polyoma) and the simian virus, SV40. They are very small viruses. There is no evidence that any human tumour is caused by any papovavirus, though genetic information related to SV40 has been detected in several human tumours. Tumours caused by the papilloma and polyoma viruses appear initially to be non-malignant and may remain so or regress; they may, however, proceed to malignancy once established. Simian virus (SV40) is one of many, that are commensal in rhesus and other monkeys, causing trouble when, as they frequently do, they appear in cultures of monkey kidney cells used for virological research or anti-viral vaccines. Though harmless to monkeys and apparently to man, they cause rapidly fatal neoplasms in baby hamsters.

The adenovirus group

The adenoviruses are similar to papova, but slightly larger. Three strains of human adenovirus, numbers 12, 18, and 31 contain transforming information and are strongly oncogenic in rodents, as are 5 of 17 simian strains. They do not, however, appear to be oncogenic in man and there is no evidence that any human tumours are derived from adenoviruses acquired from animal sources. Some adenoviruses cause chronic respiratory troubles in man, which are of long duration and persistent. Vaccines against such have been prepared from virus grown in monkey kidney cells, some of which have been contaminated with SV40 derived from the culture cells. This happened before the oncogenic properties of SV40 had been discovered; once this was known, there was widespread dismay because more than a million human beings had been treated with vaccines that were potentially infected with SV40; this was the more disconcerting, when it was discovered that the adenoviruses form hybrid recombinants with SV40. Recipients of adenovirus vaccines were traced and kept under observation; fortunately none developed cancer.

The herpesviruses

We have already encountered one instance of cancer, caused by a herpesvirus in Marek's Disease of poultry. A herpesvirus is also responsible for causing the Lucké Kidney Carcinoma in frogs, the study of which constituted an important milestone in that of viral cancers; this tumour will be considered in greater detail in a later chapter. Two herpesviruses, *Herpesvirus saimiri* and *H. ateles*, which are natural pathogens of squirrel and spider monkeys respectively, cause rapidly progressive fatal leukaemias, lymphomas and sarcomas in closely related monkey species. Man has his own resident herpesvirus in *H. simplex* type 1, which causes herpetic lesions around the mouths of young children when they first become affected, and may recur throughout life in the form of the well-known 'cold sores'. *Herpesvirus simplex* type 2 causes herpetic lesions on the genitalia and is venereally transmitted in the semen. This virus has an especial importance, because it is associated with cervical carcinoma, a serious cancer of women, of which it is believed to be the cause. It has also been suggested as a cause of carcinoma of the prostate in men. Cervical carcinoma is especially prevalent amongst women, who are promiscuous and have indulged in sexual activities at an early age. Another herpesvirus, the Epstein–Barr Virus (EBV), is the cause of glandular fever (infectious mononucleosis). The target cells of the virus are the B lymphocytes, white blood cells concerned with immune mechanisms; these cells are 'transformed' in the patient and remain transformed throughout life, even when active symptoms have subsided. The disease is, therefore, a low grade, self-limiting, leukaemia of the B lymphocytes and, though not generally recognised as such, a transmissible cancer of man. Epstein–Barr Virus is also constantly associated with the well-known Burkitt's Lymphoma, a rapidly fatal cancer of the neck and jawbones of African children; it is also associated with nasopharyngeal carcinoma, a serious cancer common in eastern Asia; there is little doubt that it is the actual cause of both these cancers. All herpesviruses, whether oncogenic or not, have properties of latency. They are large viruses, in which the core contains double-stranded DNA.

The oncornaviruses

The oncornaviruses all possess transforming ability. In birds, there are five sub-types, A, B, C, D, and E, distinguished by the particle morphology and other properties. Only B, C, and D are of importance in mammals and so potentially in man. These viruses contain RNA and multiply by budding when attached to the host cell plasma membrane. Their oncogenic properties are thought to be due to induced chemical and morphological changes in this membrane, so that cell multiplication ceases to be regulated by 'contact inhibition'. The viruses also possess a gene for the manufacture of an enzyme, known as 'reverse transcriptase', by which DNA copies of themselves are constructed. In the DNA or 'provirus' form, their genes become integrated in the host DNA, as described above. Intact viral genomes or viral genetic information can be demonstrated in virtually all mammalian cells, including human.

Type B oncornaviruses are the cause of breast tumours of mice; they are frequently present in human milk and can be found in a great many samples of breast tumour tissue in women, but not usually in normal breast tissues. There is every likelihood, that they are the actual cause of breast cancer in women, the development of which is also dependent on some hormone factors. Type C oncornaviruses are the cause of leukaemias, lymphomas and sarcomas in birds, rodents, cats, dogs, cattle, horses and other animals incuding monkeys and apes. Type C particles are widely associated with similar neoplasms in man; they are likely to be the cause, though there is as yet no definite proof.

Oncornaviruses are transmitted in three ways: 1. vertically in the sex cells, 2. horizontally shortly after birth, as with mammary carcinoma of mice, which is transmitted in the mother's milk, and 3. horizontally at a later stage after birth. With vertically transmitted virus, antibody is only weakly developed, because the foetus has acquired tolerance while in the womb.

The antigenic relationships of the oncornaviruses of different mammalian species suggest that oncornavirus DNA has been transmitted vertically generation to generation for at least 40 million years, evolving *pari passu* with the host. These relationships reflect also the taxonomic relationships of the host species. They may also suggest that certain oncornaviruses have been acquired from other species by horizontal transmission. For example Harvey Rabin (1978) states that non-human primate oncornaviruses closely resemble those of a wild-living Asiatic mouse, but are far removed from those of the house mouse; he suggests their derivation from those of the Asiatic mouse.

It will be evident from what has been written that the study of carcinogenic viruses requires knowledge of the behaviour of a virus once it has entered the host cell. Genetic information for cell transformation is to be found only in certain viral groups and is absent from the vast majority. Furthermore, the groups possessing that information, except for the papova- and adenoviruses, are not only unrelated but completely dissimilar; whether similarity exists between the genes which cause transformation is undetermined. One feature all possess in common is the ability to persist in the host cell in a state of latency, and transforming ability may well be connected with this form of relationship between virus and host.

Recent evidence suggests that the substance known as Interferon may induce regression of some malignant tumours. Interferon was isolated by Alex Isaacs from living cells and found to have a specific effect in controlling viral infections. It appears to be effective against most or all viral infections, when they are actively causing disease symptoms; it would appear to be a natural biological defence against this form of infection. However, each animal species manufactures its own Interferon, which is only effective in that animal. Progress on research into Interferon in the comparatively long interval since it was first described has been unprecedentedly slow, because of the difficulty of producing it in sufficiently large quantities for effective research. These difficulties appear likely to be overcome within the next few years; if so, there will be available for the first time a biological agent of the nature of an antibiotic capable of controlling viral infections. Its effect on the progress of malignant tumours has been described more recently, and a number of reasons has been suggested for its action in this respect mostly concerning a control of the differential rate of division between normal and transformed cells. It could, however, well be that its action is due to an inhibiting effect on viruses or viral genes responsible for causing the cancer.

In the following chapters, we shall study the history of research into the viral causes of the animal cancers over a great many years. This is not well known, in spite of the length of time during which these researches have been made and the enormous amount of information that has been acquired. In the course of this survey, it will be seen how successive advances have been dependent on the development of new technological instruments of research and the application of more advanced techniques.

References

Gross, L. (1970). 'Oncogenic Viruses', 2nd edn, 991 pp. Oxford: Pergamon Press

Kurstak, E. and Maramorosch, K. (eds) (1974). 'Viruses, Evolution and Cancer'. New York and London: Academic Press

Rabin, H. (1978). Studies in nonhuman primates with exogenous type-C and type-D oncornaviruses. *In* (Chivers, D. J. and Ford, E. H. R., eds) 'Recent Advances in Primatology'. Vol. 4. London: Academic Press

2

The Somatic Theory of Cancer

There are two main theories as to the cause of the malignant diseases, collectively known as 'cancer': 1. the somatic theory, and 2. the viral theory. Students of the cancer problem, who study first the human cancers adhere for excellent reasons to the somatic theory; those, who study first the animal cancers, for equally good reasons mostly espouse the viral theory.

As a veterinary pathologist, who has specialised in simian (non-human primate) pathology, I myself belong to the latter persuasion, and found it difficult at first to understand how infectious causes of cancer could fail to be accepted and applied to the human situation. Fortunately, the evidence for the somatic theory has been presented in a most articulate form by two outstanding scientists, Sir Macfarlane Burnet (1978) and John Cairns (1978). It is, therefore inexcusable not to be acquainted with the arguments for the somatic theory, and it is a rash man, who will cross swords with such eminent antagonists. Unfortunately, there is no succinct account of the infectious animal cancers, and it is my object in this book to give such an account, and to give the evidence for their infectious origins. In my first two chapters, I intend to discuss the two theories. The proponents of the two theories rely for their arguments on different facts and evidence, which lead to different conclusions. Neither side appears to be well informed about the evidence available to the other, and some rather surprising statements are made to demolish each other's positions. If the two sides of the controversy could be reconciled, real progress might be made; my own studies lead me to believe that they can. The virologists do not appear to attach sufficient importance to the problem of mutagenesis and errors of DNA copying and repair; there is even a tendency to regard mutagenesis and carcinogenesis as synonymous on the basis of tissue culture studies; studies of cancer genetics show clearly that they are not. The geneticists, who espouse the somatic theory, in their turn do not appear to realise that the virologists do not propose viruses as such as the cause of cancer, but the presence of viral genes in the genetic apparatus of host cells. The controversy, therefore, becomes defined as to whether the aberrant behaviour of cancer cells is due to mutations in the host cells or unusual activities on the part of normally quiescent commensal viral genes. Evidence from the animal cancers would suggest the latter, but the difficulty of finding evidence for such in human cancerous tissues suggests that viral genes are not involved.

In chapters 2 and 3, then, I attempt to present as objectively as possible the two sides of the picture. In the remainder of the book, I shall as best I can present a factual account of the infectious animal cancers, although chapter 8 will be devoted to those cancers of man for which there is some evidence of a viral origin. It may well be asked as to whether it matters, if cancers are caused by host or viral genes? It does matter tremendously. If only host genes are involved, curative procedures will depend on methods of destruction of the cancer cells as such by drugs and irradiation treatments. These are methods available today and, though reasonably successful in some kinds of cancer, are both traumatic to the patient and have limited success in other cancers. If, on the other hand, viral genes are involved, a second line of treatment might be developed to neutralise the activity of the virus. The whole subject is of a complexity unparalleled in medicine. Either way, quick results are unlikely to be achieved, but success is more probable if the two polarised points of view are reconciled.

Somatic mutation

The somatic theory, as expounded by Burnet and Cairns, states that malignant changes in cells are part of the ageing process and arise from mutations affecting the DNA sequences in the somatic cells of the body. The mutations arise during the course of cell division, and are caused by errors of copying the DNA and failure of DNA repair processes. Such mutations are continually occurring throughout life and usually lead to the death of the cell or its elimination by the body's immune mechanisms. However, as the host ages, such mutations become more frequent and the immune defences weaken. Rarely such mutations impair a cell's response to the natural mechanisms, which control its rate of proliferation; it will then become neoplastic.

The clonal nature of cancers

The bodies of multicellular animals are 'clones'. A clone is an assemblage of cells arising from a single cell precursor. An animal's body is thus a clone, since it has arisen in all its diversity from a single fertilised ovum. It follows that the parent cell contains all the genetic information necessary to develop the many functions of the differentiated tissues of the adult. At first, the ovum divides into four similar cells, then eight, then sixteen, but thereafter during embryonic development the daughter cells become differentiated in form and function. This is effected by a vast array of enzymes, which are proteins manufactured by the cell's own DNA. Every daughter cell of the parent ovum retains the genetic capacity to manufacture any or all of the tissue cells that constitute the entire organism, though they do only manufacture cells appropriate to the tissue in which they are located. It follows that the majority of a cell's genes are 'repressed', and their confinement to legitimate activities is assured by feedback mechanisms arising from the cell's own DNA. Faulty copying of this DNA could lead to irregularities of behaviour, including 'derepression' of some genes. In the cancer cell, some genes are derepressed in two ways: 1. the cells are no longer responsive to signals inhibiting cell division, and 2. the manufactured product is unsuited to the parent tissue, including certain proteins normally only found in foetal tissues, 'foeto-protein'. They have in this and possibly other ways reverted to an earlier stage of development. Tumour cells have other odd characteristics, which unfit them for the role of simple throwbacks to an earlier stage of clone development; for example, they can grow in the absence of oxygen and they are immortal, in the sense that normal tissue cells always die after a few generations of growth in tissue culture, whereas cancer cells can be maintained indefinitely. These properties are hard to explain in terms of the somatic theory.

Tumour cells are themselves said to be clones ('monoclonal'), because all malignant tumours are derived from a single 'rogue' cell, in which a mutation of the type described has arisen. This is a surprising feature of malignancy, and one of the strongest arguments advanced for the somatic theory; if a virus were involved, so it is said, a number of cells in the tissue would become involved at the same time. This has been determined in the following way. During foetal development of the female, the paired sex-determining X chromosomes become differentiated; one, the paternal or maternal, becomes residual and moves to the side of the cell as a small knot of chromatin; the other remains unchanged and active. The two chromosomes are randomly selected for the role they are to play, so that in this respect all tissues are 'mosaics'. The X chromosomes can be distinguished from each other in some subjects by differences in an enzyme, known as G6PD, produced by the X chromosomes. By this means, it has been determined in the case of many of the cancers, that the tumour cells contain either paternal or maternal X chromosomes, but not both. This is taken to indicate that the cancer is derived from a single cell, that it is 'monoclonal'; the validity of such a conclusion could be argued, but other evidence suggests that it may be true.

Mutagenesis

Cancers are basically due to cell mutations, that is errors of DNA copying at the time of cell division and failure of the DNA repair process. These errors and failures may result from exposure of cells to mutagenic influences, such as various chemicals, exposure to various forms of radiation, or possibly to viral infections. Such influences are for the most part of lesser importance, since there occur a great many more mutations of somatic cells than can be accounted for by exposure to these risks. Amongst the hundreds of millions of cell divisions that occur daily, it is not to be supposed that some errors will not occur. Usually, the cell will die or be eliminated by the body defences, or the DNA repair processes will remedy the defect. However, the tendency of cells to mutate becomes more pronounced as the clone grows older, the immune defences become less efficient, and the processes of repair less effective. This is well attested in tissue cultures of cells as well as in the living body. It is such errors of copying and failure of the repair and immune processes which, according to the proponents of the somatic theory, lead to neoplasia.

If this is to be understood, a brief digression into the nature and structure of DNA becomes necessary, although this is repeated from many other books on allied subjects. DNA is a very complex molecule consisting of a core of two filaments, which form the well-known double helix. The filaments are composed of elements of the 5 carbon sugar ribose combined with phosphorus and longitudinally linked. They are transversely linked by two series of nucleotides which are complementary to each other, the same pairs always being sited in opposite positions. These paired nucleotides form a double chain along the length of the ribose filaments. The nucleotides are: adenine (A), cytosine (C), guanine (G) and thymine (T). Adenine is always paired with thymine, and cytosine with guanine. These substances are complex ring structures with a nitrogen component, known as purines and the allied pyrimidines. A chain of some 100 000 nucleotides constitue a single gene and provide the information to construct the amino acids necessary to build the appropriate polypeptide chains for manufacture of the required protein, which may be a structural element or an enzyme. The protein constructed is determined by the order in which the nucleotides are arranged. They form a kind of alphabet. Whereas an alphabet of 26 letters can be used to construct innumerable numbers of words, this can also be achieved by a series of dots and dashes as used in the morse code. The ACGT system is a form of biological morse, which can direct the construction of a very large number of proteins, the ordering of which leads to the building of a plant or animal from tissue cells devised for a specific function; that function is, furthermore, directed and controlled by the enzymes produced.

When the cell divides, the DNA filaments split down the middle to form four instead of two. Each filament then acquires a new set of nucleotides to complement those that are missing, so that the correct sequence is replaced. Cell division is very active throughout life in organs such as the skin and bowel, but occurs less frequently in other organs and does not occur at all in nerve cells. When proteins are manufactured, the appropriate segment of the double helix becomes straightened and an equivalent segment of single stranded RNA is hived off bearing the nucleotide sequences of its parent DNA. This RNA has a limited life and is known as 'messenger RNA' (mRNA). It passes from the nucleus to the cytoplasm probably by way of the nucleolus, from which it is collected by the ribosomes. In these 'organelles' it becomes attached to a segment of double-stranded RNA, which is stimulated to arrange a chain of polypeptides from the correct sequence of amino acids required to construct the protein demanded by the enzymes. Some of the enzymes constructed will be of a feedback nature, controlling the activities of the genes, including repression and derepression of their activities. The process, as described, may seem simple, but in it hundreds of both genes and enzymes are evolved, so that it is in fact immensely complex. Furthermore, errors of copying may lead to serious defects in the tissues to be reproduced. It would be surprising, if some errors did not occur from time to time, and indeed they do. When such occur, the affected cell is usually unable to reproduce further and dies. However, elaborate gene mechanisms exist, involving yet further enzymes, whereby defective DNA is repaired to its normal state. It follows that abnormal cells may arise either from defective copying or from errors of the repair process. It is from these circumstances that abnormal cells, including cancer cells, sometimes appear in a tissue. In cancer cells, there is a failure of the 'repression' mechanisms, which limit the cell's power to proliferate and produce abnormal tissues. There is also a failure of the body's immune mechanisms, which should recognise the mutated cells as alien and destroy them.

These results may occur in the natural course of events as an unfortunate accident, because of hormonal defects as for example in breast cancers, as part of the ageing process, or because of the influence of carcinogenic agents, harmful chemicals or radiation. It is admitted by the protagonists of the somatic theory that viruses may on occasion be carcinogenic, but they are regarded only as one amongst a number of others.

A difficulty of research arises from the time factor, since a great many years may elapse between exposure to a mutagenic or carcinogenic agent and development of the resultant cancer. For example, 15–20 years of heavy smoking may precede development of lung cancer, so that cause and effect are not immediately apparent. It is postulated that there are two phases of cancer, namely a pre-cancerous phase when cells are changed, and the subsequent phase when proliferative and invasive properties become developed. It is also suggested that, before cells become neoplastic, two mutagenic influences are involved, one of which may be genetic. Thus, cells with a genetic defect exposed to a mutagenic agent or involved in an error of copying could become neoplastic; alternatively, successive exposure to two mutagenic agents could lead to carcinogenesis. This is written 'g + s' or 's_1 + s_2'; that is 'genetic + somatic' or 'somatic$_1$ + somatic$_2$' mutagenic agents.

The cell, of course, is engaged in numerous activities apart from those of DNA copying and protein synthesis. However, these are the factors believed to be concerned, first with mutagenesis or cell mutation, and secondly with carcinogenesis. The numbers of cells, genes and enzymes involved in cell divisions is vast, running into billions daily. Somatic mutations are continually occurring and may not always be the result of errors, since they may play some part in natural and important processes. For example, it is believed that some relocation of nucleotide sequences is essential to enable the body to build the millions of antibodies required of it in response to infections and the presence of alien protein and other chemical substances. Most mutations are, however, probably the result of accidental transfers or linkages of nucleotides or else of faulty construction of enzymes or proteins. The activation ('derepression') of genes depends on the presence of numerous enzymes, produced by other genes, which control the system. Other enzymes, produced by yet other genes, are responsible for inhibiting the activity ('repressing') genes, whose products are not required. Yet other genes producing still more enzymes initiate their repair process when faulty copying has occurred. A number of mutations, which are relatively harmless, probably persist, but may in aggregate over the years be responsible for the ageing process. A few mutations may confer some benefit, either temporarily or permanently, and give greater survival advantages. It is such rare mutations, that are believed to be responsible for evolutionary changes which become established over many thousands of generations.

The natural mutations are continually occurring and are far more numerous in any ordinary population than those that are induced by external carcinogenic factors. The latter have, however, become increasingly important, as man's way of life becomes increasingly sophisticated. Carcinogens cause characteristic changes in nucleotide sequences, which vary with but are specific to the carcinogen involved; the repair processes also vary with the nature of the damage inflicted. The somatic theorists of cancer origins find that natural or induced errors of DNA copying and repair are more than adequate to explain its causes. They see no reason to look further for such causes, since prolonged and very expensive searches for viral involvement in human cancers have been mostly unrewarding. However, it is conceded that in certain cancers, such as Burkitt's Lymphoma and possibly Hodgkin's Disease, some viruses may be acting as mutagens or even carcinogens. Even so, the viral involvement is likely to be due to some impairment of the immune system, which may be genetic, due to its being overtaxed by some other stress, or due to pre-cancerous mutagenic change. From the aspect of the human cancers, these arguments appear to be valid in the absence of further evidence. They appear, however, to be contradicted by the evidence from animal cancers, and we must review the objections put forward to the viral theory.

Objections to the viral theory

If all animal cancers were caused by virus infections, whether of complete replicating virus or of viral genes insinuated into the host genome, it would be difficult to suppose that man was the only exception to prove the rule. If it could be shown that, contrary to the evidence for a viral cause of most animal cancers, the viral presence was an artifact or that the viruses present were mere opportunists able to colonise the mutated cells, then the somatic theory would prevail. No such evidence has so far been found. However, the objections to viruses as causes of animal cancers rest on the following points: 1. the epidemiology of cancer in man, and 2. that the infectious animal cancers are in fact artifacts created in the laboratory.

(*i*) *Cancer epidemiology.* The epidemiology of cancer in man supports the somatic theory. The more important cancers are remarkably stable in their incidence, as shown by annual mortality statistics. Their frequency also increases in linear progression, when logarithmically plotted, with advancing years. In both Europe and the USA, the incidence of lung cancer has been on the increase during the past thirty years, while there has been a corresponding decrease in the incidence of cancer of the stomach; the total cancer incidence has remained stable, in spite of this. The lung cancer increase is attributed to smoking habits, and there is little doubt that tobacco tars are carcinogenic, though the majority of persons, who smoke, do so with impunity. The reasons for the reduction of stomach cancer incidence are unknown. It is further stressed that only rarely do cancers occur in 'clusters'. This is to say, that although there is an undoubted familial susceptibility, epidemics amongst persons, who associate together, are very rare except where they are exposed, as in industrial complexes, to similar carcinogenic influences. Were infectious agents involved, such epidemics would be expected.

The stable annual incidence associated with Western ways of life and the age association strongly support the belief that cancer is associated with constant and predictable errors during cell division and age-linked diminution of immune responses. Some cancers, such as the leukaemias and cancer of the breast in women admittedly occur quite frequently in quite young people, but could result from genetic causes, from exposure to carcinogens or even from unfortunate mutations occurring at a time of immunological embarrassment. Burnet stresses that the epidemiological picture is the reverse of what would be expected if cancer resulted from viral infections; if this were true, the incidence would be higher in young people and would diminish with age. In cancers, such as Burkitt's Lymphoma, in which viral involvement can be expected, the disease occurs in young people as would be supposed. In any case, so he argues, the geographical distribution of Burkitt's Lymphoma and the evident involvement of a secondary factor, probably malaria infection, suggest that viral involvement may not be of primary importance.

The transmissible cancers of chickens and mice can also reveal an age-associated incidence, if allowance is made for the natural longevity of the species involved. Therefore, it is argued, as with Burkitt's Lymphoma, that the involvement of the virus may only be of secondary importance. These cancers only occur with regularity in artificially bred strains, which shows that genetic and somatic factors are likely to be of more importance than infectious. Thus the viral involvement must be regarded as an artifact created in the laboratory.

(*ii*) *Carcinogenic viruses as laboratory artifacts.* The infectious cancers of chickens, mice and other animals only occur with regularity in domestically or artificially bred strains, though it is difficult to explain away the flock or herd incidence of Marek's Disease, the avian leukosis complex or bovine papillomas. Even in these, however, Burnet's age incidence has validity, since their occurrence is in young animals, which may serve to differentiate them from the human cancers. Cairns goes so far as to suggest, that the virus of Rous Sarcoma of chickens, accepted for the past seventy years as the cause of the disease, may be an artifact, because it has acquired a mutated gene from an original host by gene exchange. It is known that viruses and host cells do exchange genes in this way, so that this argument, though seemingly specious and somewhat irrelevant, cannot be invalidated. To the poultry owner, it is important that the virus causes disease in his birds; it does not matter to him, whether it is a host gene or a viral gene that is responsible.

Neither Burnet nor Cairns consider the evidence derived from the transmissible simian cancers, which are discussed in fair detail in a later chapter. Nor do they consider the problems arising from the presence of viral genes of known oncogenicity, which are commensal in the host genome. Their significance will also be considered in the chapters that are to follow. Meanwhile, having stated the case for the somatic theory, it is necessary to do the same for the viral theory.

References

Burnet, Sir M. (1978). 'Endurance of Life'. Melbourne University Press
Cairns, J. (1978). 'Cancer, Sciences and Society'. San Francisco: Freeman

3
The Viral Theory of Cancer

In 1964, the US Congress decided to fund a long-term project entitled the 'Virus Cancer Program'. It was at that time widely believed that cancer was caused by viruses, and that with adequate financial support its causes could be determined. If so, methods of prevention and cure could surely be devised. The program was abandoned in 1979 after a staggering $549 million had been spent on it. It might be supposed that such a massive effort over a period of twelve years could not fail to give a positive answer, and that if it failed to do so it could safely be assumed that viruses were not implicated. The program has predictably been widely criticised as an enormous waste of resources, which could have been more profitably employed in other directions. The program was started with high hopes and a note of cautious optimism pervades the earlier annual reports. It became clear, however, that the search was not for viruses as such but for viral genes capable of manipulating the genetic system of the host. Latterly the program seems to have lost sight of its initial objectives, and a disproportionate amount of effort was devoted to the identification of viral genes in normal and cancerous tissues.

On the face of it, the program would appear to have been an expensive failure in that proof of a viral origin of the commoner human cancers was not obtained. On the other hand, no proof was forthcoming that human cancers are not caused by viral infection and circumstantial evidence strongly suggests that they are. The protagonists of the viral theory would claim that viral causes have been adequately proved for all animal cancers that have been sufficiently studied. Whether this is so, in the light of claims by the somatic theorists that they are in fact artifacts, may be judged by the reader from the accounts given in the following chapters.

Nobody will deny that the program achieved great advances in the field of virology, especially in the techniques of handling and studying viruses, in virus/host relationships, and in viral genetics and immunology. A potentially important discovery was that of the influence of Interferon on the course of cancer development. Interferon was tested on tumours in the belief that it was a specifically anti-viral agent, and that if it showed activity this was because of its effect on a virus that was causing the cancer. It is now postulated by the somatic theorists that Interferon has some effect on the somatic genome of the cancer cell.

Perhaps, the most significant discovery made during the course of the program was that viral genes exist as commensals in the reproductive and somatic cells of all vertebrates studied, including man. The genes belong to a viral group the oncornaviruses, a great many of which have oncogenic potential and are the known cause of important animal cancers. The viral genes behave as if they were host genes, dividing *pari passu* with them, being transmitted vertically from parent to offspring, and evolving over hundreds of millions of years in conformity with the host, to an extent that their properties reflect evolutionary relationships. Equally surprising was the discovery that another group of potentially oncogenic viruses, the herpesviruses, also reflect host relationships over equally long evolutionary periods, though they are acquired shortly after birth not being transmitted through the fertilised ovum. Genetic information of potentially oncogenic viruses is, thus, omnipresent in the somatic cells of vertebrates, man included. The viral theory claims that it is the influence of these genes, which causes the aberrant behaviour of the host cell rather than anomalous reactions of the host cell genes. It is stressed that cancer cells possess certain properties, such as the ability to proliferate in the absence of oxygen and immortality, which are not present in the original clone and must be acquired by extraneous genetic input. It was established during the course of the program that viral genes can dislodge some of those of the host and replace them in the DNA filament. Mutagenic and carcinogenic influences could, therefore, act on resident viral genes rather then on host genes; they could be responsible for errors of DNA copying and repair. In this way, the picture displayed by the somatic theory is not in its essentials at variance with the viral theory. However, if the viral theory is correct, an additional avenue of attack on the cancer problem becomes available.

The oncogenic viruses

A number of viruses can cause reversible, self-limiting or benign tumours resulting from cell proliferation, chiefly in the skin. Only four groups are implicated in causing neoplastic changes in tissues. Two of these groups are closely related, the papovavirus and adenovirus groups; the other two, the oncornaviruses and the herpesviruses are quite dissimilar, morphologically, genetically and immunologically. In addition, not all members of each group are oncogenic. All, however, possess one common factor, namely that they can lie dormant in the host cells for long periods or throughout life without causing disease symptoms or cancer.

A further feature of these viruses lies in the host response, which varies with age and species. A number of the viruses only cause cancer, if very young animals become infected and then onset of cancer may be delayed for a prolonged period. With some viruses, infection of the natural host may be symptomless or very mild, while other species develop ferbrile or haemorrhagic types of disease and others cancer. Even in the natural host, infection in the very young may cause a febrile or haemorrhagic type of disease and neoplastic in older. Such differences of pathogenesis can only be attributed to variations in host response, which could be genetic or immunological. Here again we must note the similarity to the somatic theory, which postulates either genetic or immunological insufficiency as predisposing factors in the development of neoplasia. Whereas the somatic theory supposes consequent errors of DNA copying, the viral theory supposes that these factors will lead to abnormal activity of commensal viral genes.

The transmission of oncogenic viruses

Transmission of oncogenic viruses may be 'horizontal' or 'vertical'.

When viruses are transmitted horizontally, they are passed in the same way as any other infectious organism from an infected animal to an uninfected. However, further elaboration is needed, because there are three ways in which the infection may pass with important differences.

In its simplest form, the oncogenic virus behaves like any other infectious virus. It is acquired by contact transmission, aerosol or infected discharges. It then invades its target tissues, proliferates to form new infectious viral particles, which are shed to infect further susceptible hosts. In the normal course, the virus would succumb to the host's immunological defences and be eliminated. The oncogenic viruses, however, seek alternative target tissues, where they are less exposed to the defences; although proliferation of viral particles is prevented, viral gene activity causes somatic mutation and carcinogenesis of the host cells. A typical instance is Marek's Disease of chickens, a lympho-proliferative disease which causes serious losses in infected flocks. The virus is one of the herpes group, whose primary target organ is the feather follicles in the skin. Active infection of the follicles occurs and viral particles are shed in the feather dander and are infective to other birds. However, the virus also enters cells of the lymphatic system, which are stimulated to uncontrolled replication. This disease is certainly no artifact. It is, indeed, controlled by vaccines, which are commercially available. The vaccines are made from live virus, which has been modified by growth in tissue culture and lost its oncogenicity, or else from a related virus, the natural host of which is turkeys and which infects chickens without causing neoplasia. Strangely, these vaccines permit super-infection of the chickens with the virus of Marek's Disease, but its carcinogenic activities are inhibited.

Another instance of the classical form of horizontal viral transmission is seen in cat leukaemia, which passes horizontally from cat to cat by inhalation or in the nasal discharges. It is caused by an oncornavirus, feline leukaemia virus (FeLV). This is a straightforward naturally occurring infectious disease of cats, certainly no laboratory artifact. If we seeks a transmissible cancer in a truly wild animal, we may draw attention to the Lucké Renal Carcinoma of the American leopard frog (*Rana pipiens*), caused by a horizontally transmitted herpesvirus. The affected frog populations live in wild undisturbed ecological situations.

The second form of horizontal transmission to be considered is characteristic of the normal commensal herpesviruses, which are ubiquitous in their distribution throughout the animal kingdom and as far down the evolutionary scale as the Fungi. The newborn animal is uninfected, but acquires infection from the mother shortly after birth. There is a transitory period in which mild herpetic lesions may cause some distress, but they quickly subside and the young animal becomes a lifelong carrier of the virus. Under certain conditions, herpetic lesions may reappear and on rare occasions serious diseases may be caused by the virus. Though a number of herpesviruses become commensal, comparatively few possess the power to cause cancer. Man's own chief commensal herpesvirus is *Herpesvirus simplex*, of which there are two strains (HSV_1 and HSV_2). HSV_2 is transmitted venereally during coitus causing herpetic lesions on the genitalia of women; there is strong evidence that this virus is the cause of the common cervical carcinoma of women. The virus closely resembles HSV_1, from which it is believed to be a derivative. HSV_1 is transmitted in the usual way shortly after birth and causes herpes lesions on the mouth and lips. There also exists a venereal tumour of dogs, caused by a herpesvirus.

Herpesviruses, acquired at birth, are usually harmless to their natural hosts. There is little evidence that they cause cancer in them, though they might be implicated in some human cancers, such as Hodgkin's Disease and those of the mouth and pharynx. When transmitted to unnatural hosts, they often cause very serious diseases, sometimes neoplastic. The classical instance is that of the squirrel monkey herpesvirus (*H. saimiri*), which causes by horizontal contact transmission a rapidly fatal form of lymphoma in the related New World marmosets and owl monkeys. The natural herpesvirus of Asian macaque monkeys causes a fatal encephalo-myelitis in man, though not cancer. These instances again cannot be described as artifacts.

The third method of horizontal transmission of oncogenic viruses is that seen in the mammary tumours of mice, caused by an oncornavirus. The agent, known as the milk factor, is transmitted to the newborn mouse in the mother's milk; it causes mammary carcinoma in the female mice, after they come of breeding age. Females can also be infected by transmission of virus in the semen of the male mouse, which shows no clinical signs of infection. It is evident that some endocrine factor is involved in this condition; there are also genetic influences, since some laboratory bred strains of mice are resistant, while others show varying degrees of susceptibility. Wild mice occasionally suffer from the condition, and it was by breeding from susceptible and resistant strains that made it possible to determine the cause. It is this kind of situation, which is regarded by the somatic theorists as 'laboratory-made artifact', but this is difficult to support when they themselves admit genetic factors as amongst the influences leading to cancer. There is good reason to believe that some human mammary cancers may be due to a similar cause, as will be discussed in a later chapter.

In a sense, transmission of normally commensal, but potentially oncogenic viruses to the newborn whether by contact with the mother or through the milk, is a form of vertical transmission, though there is an important difference discussed later. True vertical transmission is confined to the oncornaviruses, since it is only they, which are congenitally passed by way of the foetus. These viruses are extremely difficult or impossible to detect, except when provoked to replicative or antigenic activity. They do, however, at certain times produce entire viral particles, which can be identified. Kalter and his colleagues at the South West Foundation at San Antonio in Texas were able to identify these viruses in baboon placentas at the time of parturition and subsequently in ova and undifferentiated embryos. It was subsequently shown that similar viral particles could frequently be demonstrated also in human embryos and placentas. It has now been established that oncornavirus genes are almost universally present in the genome of most vertebrate species, including man. In spite of the potential danger, that they may be associated with cancer, it is thought that they may play a mostly beneficial role, like commensals of other groups of microorganisms, in promoting evolutionary mutations. Their transmission is vertical from parent to offspring, and they may be acquired from the male or female reproductive cells, and so presumably be either homozygous or heterozygous. Normally they remain inactive, but can be provoked by various means either to replicate as entire viral particles or to oncogenic activity. Since they are RNA viruses, replication takes place in the cell cytoplasm, where unlike other viruses their presence does not lead necessarily to destruction of the host cell. The evidence that it is they which cause cancers rather than faulty copying of the DNA of somatic cells comes from studies of the genetic apparatus of oncogenic viruses.

The genetics of oncogenic viruses

Oncogenic viruses carry an 'oncogene', that is a special gene which causes cancerous changes in host cells. This gene is located on the DNA filament in a position that is separate from those genes concerned with replication, as has been shown by some very ingenious studies. Viral genes, like those of more advanced organisms, are subject to mutations, the incidence of which is more or less predictable. One such mutation is the 'temperature sensitive' (ts). Cells carrying ts mutants can be harvested in tissue cultures. When grown at 32°C the virus both replicates and transforms cells to a neoplastic condition. However, when grown at 39.5°C, the virus loses the property of replication as complete virus, but is still able to cause cell transformation as a result of its location in the host genome. This discovery is of importance to the viral theory. It shows that the power to transform cells to the neoplastic condition resides in certain specialised genes and there is no evidence that such genes reside in vertebrate DNA, unless they are of viral origin. Temperature-sensitive mutants could indeed be involved in cancer causation in inflamed tissues, if their temperature was raised.

Genetically, viruses are very simple organisms compared with other forms of life. Mature viral particles vary greatly in size according to the virus group, to which they belong. They may be comparatively large, indeed sub-microscopical as with the pox group, or very small as with the papovaviruses, some of which are oncogenic. They possess a coat, constructed of protein and other substances, which is of characteristic shape, structure and complexity. The viral coat is shed, when a host cell is penetrated and the contained substances, consisting largely of viral DNA, become free within the cell. The RNA viruses remain within the cytoplasm, attach themselves to the cell membranes and replicate by budding new particles. The DNA viruses penetrate the nucleus, where the viral genes manipulate the host DNA to produce mRNA, which will stimulate the ribosomes to construct new virus. The viral genes are capable of some activity independent of the host cell. For example, the myxovirus of influenza constructs an enzyme 'neuraminidase', which has the property of breaking down the mucous coating of the cells of the respiratory tract, so facilitating the entrance of the virus into them. Similarly, RNA viruses, including the oncornaviruses, produce an enzyme of the polymerase group, known as 'reverse transcriptase' by which DNA copies of their RNA genes are constructed able to penetrate the cell nucleus. These viruses, thus, have the best of both worlds; they can replicate in the cytoplasm as purely RNA viruses, or enter the host genome and penetrate the DNA, where they reside and behave like host genes, until they are again able to reform as complete RNA virus. Their genes will be subject to 'repression' mechanisms which apply to the host cell, but in the event of 'derepression' they are free to return to the cytoplasm and replicate. In the event of partial 'derepression' affecting only the oncogenes, conditions could be created in which cancer would result.

Viral DNA is a continuous filament coiled in a helix with nucleotide sequences similar to that of prokaryotic and eukaryotic organisms, except that it is much shorter and simpler and so of lower molecular weight. The numbers of proteins, which could be coded by the DNA of some viruses has been calculated, being as low as 10–12 for the papovaviruses and 15–25 for the slightly larger adenoviruses. In such simple life forms, it should be possible to map the position of the genes on the DNA filament and to determine their function. Some success has been achieved in this, but there are difficulties. Genes can only be recognised by their products, which in viruses may be enzymes or antigens. Any abnormal products found in virus infected cells may be reasonably supposed as due to viral activity. The difficulty arises in determining, whether such substances are the direct product of viral genes or abnormal host cell products resulting from the viral takeover. In solving this problem, again advantage is taken of various viral mutations which may inhibit the activity of certain viral genes, thus enabling their function and relationship with other genes to be determined.

Before viral DNA construction is started by the manipulated host genes, there appear a series of 'early antigens', which are probably the product of viral gene activity and an intermediate stage in the building of complete viral particles. However, it is now recognised that they are preceded by yet earlier antigens, of which the neuraminidases and reverse transcriptases, already mentioned, are examples. They have been given the cumbersome name of 'pre-early antigens'. Amongst them are enzymes, which remove certain segments of host cell DNA and replace them with viral DNA. It is in this way that viral control of the cell is achieved. Viruses can also exchange genes with the host and with other related viruses to form hybrids or 'recombinants'. During the process of host DNA penetration, some viruses lose important genes, including those which stimulate replication; in this way they become resident in the DNA, but any activity, including oncogenesis, is inhibited. Instances of 'incomplete' viruses are those which cause avian and feline sarcomas. Only when the host cell is super-infected with a 'helper' virus, usually the related leukaemia virus, does cancer develop; the cancer is a sarcoma, and development of leukaemia from the leukaemia virus is inhibited.

The recombination of viral genes may be of importance, because the hybrids acquire new properties. For example, a virus with the property of invasiveness might acquire an oncogene and thus become oncogenic. Or, a resident and quiescent virus possessing an oncogene might acquire a gene for invasiveness. These possibilities add an additional explanation of the ways in which cancers may, in some circumstances, become developed, of which the somatic theory takes no account. It would seem that once the cell is entered, the viral genes become 'foot-loose', to use Sir Christopher Andrewes' term. Probably, different genes penetrate different sites in the host genome, thus diversifying the viral manipulation of the cell. Some genes may be discarded because they cannot find a niche, thus producing the 'incomplete' virus; it is even possible that only the oncogene could be retained, available for acquisition by some other viral invader. However, when all the genes remain intact, active-free virus may be reconstituted, though this option remains 'repressed', until favourable conditions are created for viral replication.

If these facts and surmises are correct, there exists the possibility that complementary viruses could be found that would either suppress or remove the oncogene, as evidently happens with the Marek's Disease vaccines. In those rare cases of recovery from terminal cancer, it is possible that some accidental infection with some such virus has had just this effect.

Conclusion

The viral theory of cancer rests its case on three main premises: 1. that all animal, as opposed to human, cancers that have been intensively studied are in one way or another consequent on viral intervention; it is inherently improbable that human cancers alone should have a different etiology; 2. that cancer cells possess properties, which vertebrate genes lack the genetic information to contrive, and 3. that virtually all vertebrate cells, reproductive and somatic, human and animal, harbour commensal viral genes in their genome, which have the proved ability to cause carcinogenic changes.

Virologists distinguish between those cancers caused by *exogenous* and *endogenous* viruses. Exogenous viruses, those acquired from external sources, present a different epidemiological picture to the endogenous, or resident, viruses. Marek's Disease is typical of what occurs with an exogenous viral infection. If introduced into a poultry flock, it spreads rapidly and causes serious losses in typical epidemic fashion; there is no particular age association. Cancers due to endogenous viruses have the age association and sporadic incidence, to which attention was drawn in relation to human cancers in chapter 2. The resident viral genes are susceptible to the same influences as those that are postulated as responsible for errors of DNA copying in somatic cells. The only real question lies in whether alien viral genes are responsible or the host's own genes in somatic cells. The difference is not academic, because new horizons for cancer control would be opened, if the viral theory were correct.

The pages that follow give a historical account of the animal and some human cancers. It cannot be claimed with certainty that all animal cancers are caused by viruses, but it is difficult to avoid that conclusion on the evidence. The claim that involvement of viruses in animal cancers is due to artifacts may be judged by the reader.

4
The Transmissible Tumours of Fowls and the Lucké Tumour of Frogs

I Introduction

At the turn of the century, the infectious nature of disease had aroused widespread interest. Louis Pasteur, in 1880, published his results on the infectious nature of fowl cholera, from which some 10% of all poultry in France died annually. He not only demonstrated and isolated the causal organism, *Pasteurella multocida,* but by growing the organism in culture was able to produce harmless attenuated strains, which protected susceptible birds injected with it. His work on rabies extended the scope of his discoveries to a disease caused by an ultra-microscopic organism, which he was at that time unable to isolate and identify. He demonstrated further that it was possible to protect dogs not only before but after they had been bitten by a rabid animal. On 6 July, 1885 he successfully treated a boy, who had been bitten by a rabid dog. Meanwhile, other scientists were also contributing to microbiology, notably Robert Koch. In 1876, he had isolated the *Bacillus anthracis* and proved it to be the cause of anthrax. In 1882, he discovered the tubercle bacillus, *Mycobacterium tuberculosis.* In twelve years, Koch and his pupils discovered the specific organisms, which are the cause of eleven diseases. He was latterly, from 1897 to 1906, studying such diseases as malaria in New Guinea, rinderpest in South Africa, plague in India and sleeping sickness in Uganda. Meanwhile, Iwanowski (1894, 1899) had demonstrated for the first time the existence of an ultra-microscopic virus, tobacco mosaic, and proved it to be the cause of mosaic disease in tobacco plants. In 1898, rabbit myxomatosis was shown to be caused by a filterable virus by G. Sanarelli in Montevideo, Uruguay, and in the same year Loeffler and Frosch in Berlin showed that a filterable virus was the cause of foot and mouth disease in cattle. Proof of the viral origins of other diseases, such as smallpox, rabies and herpes followed.

In spite of successes with other forms of disease, similar successes were not being achieved with filtered material from cases of cancer. Researches had, therefore, been mostly confined to the transmission of cancers from animal to animal by transplanting fragments of tumours under the skin or in organs. Such tumour cells might be rejected, but, if accepted, an identical tumour developed in the recipient. Novinsky (1876) in Russia removed a carcinoma from the nose of a dog and transplanted fragments of tumour tissue under the skin of several other dogs, one of which developed an identical tumour together with metastases in the neighbouring lymph glands. Wehr (1888) also reported that he had successfully transmitted a tumour from dog to dog in 1884. Hanau (1889) transmitted a rat tumour by injecting fragments into the testes of two other rats. Similar results were reported by von Eiselsberg (1890) in Vienna. Morau (1891) transmitted mouse mammary carcinoma through successive passages from mouse to mouse. Firket (1892) and Velich (1898) both transmitted rat sarcomas through a number of passages. From that date on, tumour transfers became more widespread and an accepted method of studying cancers.

Transplantation of tumours was not, however, uniformly successful and certain facts emerged during the course of these experiments. Initially, blood relatives of the original host might have to be employed; younger hosts or even newborn animals would give more consistent results; certain sites of inoculation were more successful than others. However, with successive transplantations some tumours became more potent and could be transplanted with less difficulty. Eventually cell lines of malignant tumours were developed, which could be grown with regularity. Even so, the development of inbred strains of mice facilitated such studies. As a result of these experiments, the impression grew that cancers were strictly host specific, which in relation to the oncogenic viruses was later to be found erroneous.

In the light of this work, the transmission of chicken leukaemia by Ellerman and Bang (1908) and of chicken sarcoma by Rous (1911) by means of filtered cell-free material were unexpected results, and further developments were slow in coming. Just how slow is shown in Table 1, which gives the more important dates in the study of filterable viruses as the cause of cancer up to 1961.

Since 1966 many other viruses have been studied, which cause cancers in a number of hosts, including primates. There was an interval of 43 years between the attribution of chicken leukaemia to a viral cause and that of mice. This is a measure of the enormous difficulties attendant on these researches.

At the time they made their discovery, Ellerman and Bang were not to know that the chicken leukaemia they studied was but one of a complex group of neoplastic diseases, which has taken a great many years to unravel. Rous, too, though he realised that more than one virus was likely to be involved, could not guess at the complexity of the problem. The chicken leukaemias are today grouped as the chicken leukosis complex, a complex of

Table 1 Some of the more important oncogenic viruses. (After Gross, 1970)

Date	Virus	Reference
1908	Chicken leukaemia	Ellerman & Bang (1908)
1911	Chicken sarcoma	Rous (1911)
1932	Rabbit fibroma	Shope (1932)
1933	Rabbit papilloma	Shope (1933)
1934	Frog kidney carcinoma	Lucké (1934)
1936	Mouse mammary carcinoma	Bittner (1936)
1951	Mouse leukaemia	Gross (1951)
1953	Mouse parotid tumour	Gross (1953)
1957	*ditto* (i.e. polyoma) virus	Stewart *et al.* (1957)
1960	Vacuolating simian virus 40	Sweet & Hilleman (1960)
1962	Vacuolating simian virus 40	Eddy *et al.* (1962); Girardi *et al.* (1962)
1962	Human adenovirus type 12	Trentin *et al.* (1962)
1964	Cat leukaemia	Jarrett *et al.* (1964)
1966	Mouse osteo-sarcoma	Finkel *et al.* (1966)

diseases probably caused by a number of related, but not identical, viruses. Those transmitted by Ellerman and Bang were the ones known as erythroblastosis and myeloblastosis, which affect the parent red and white blood cells in the bone marrow. Many details concerning the various diseases and the viruses, which cause them are still not clear; newer knowledge suggests that the related viruses may exchange genes, so that they become less readily identifiable. A tentative classification of the complex is given in Table 2.

Table 2 Chicken leukosis complex

1 *Leukaemia group*
Erythroblastosis and myeloblastosis. The viruses attack the cells of the bone marrow, parent to the red and white blood cells respectively. Transformed cells are present in the circulating blood.

2 *Extravascular group*
Diseases of this group are known as 'aleukaemic', because transformed cells are rare in or absent from the circulating blood.

(*i*) *Lymphoid leukosis.* (visceral lymphomatosis or big liver disease).
A disseminated lympho-sarcomatosis with infiltration of the liver, lungs and other organs with lymphoid cells.

(*ii*) *Ocular lymphomatosis* (grey eye disease), in which there is infiltration of the eyes with lymphoid cells.

(*iii*) *Osteopetrosis*, a disease of the skeleton, in which infiltration of the periosteum (the germinal membrane covering the bones) with lymphoid cells leads to overgrowth of bone tissue.

(*iv*) *A variety of other tumours*, e.g. sarcomas, nephroblastomas (kidney tumours) and haemangiomas (blood vessel tumours).

These tumours are all caused by related viruses of the oncornavirus group, and two or more forms may co-exist; indeed, the virus of one form may induce activity in a related virus and thus induce an alien form of tumour. Filtrates of tumours sometimes cause sarcomas only, or sarcomas together with leukaemia, or adenocarcinoma of the kidneys. Young birds of certain breeds are more susceptible to infection than others, and the infection may be passed vertically through the eggs or horizontally by contact with infected birds. The 'target' cells of these viruses are the blood cells and their precursors, together with the reticular cells which line the circulatory system.

Another neoplastic disease of chickens was formerly thought to belong to the leukosis complex, namely neural lymphomatosis (fowl or range paralysis, or Marek's Disease), in which nerve cells become infiltrated with lymphoid cells and paralysis results. This disease is now known to be caused by one of the herpesviruses and is no longer classified with the leukosis complex.

II The sarcoma of chickens

Tumours of the Rous Chicken Sarcoma Complex are also caused by a variety of oncornaviruses. Rous (1911) described the transmission of one of them by filtrate; he studied in all, together with his colleagues at the Rockefeller Institute in New York, (Rous and Murphy, 1913, 1914) 45 tumours, to each of which identifying numbers were given and the morphology and histology of each was described in detail. Although some of the tumours were identical, others possessed distinctive characters. Rous believed that several related viruses were involved, although the same virus could on occasion induce different types of tumour. Rous' viruses are more potent in chicks than in adults and there is a breed susceptibility. Newly hatched chicks, on the other hand, develop a haemorrhagic disease, not tumours. This variability in the type of disease caused is a feature of oncogenic viruses, which may cause a febrile type of disease in one animal species and tumours in another.

Of the 45 chicken tumours, studied by Rous and his colleagues, only three yielded infectious virus. The original number 1, a spindle-cell sarcoma from the right breast of a Plymouth Rock hen, had been originally transmitted in series by implantation of tumour fragments; successful transmission by filtrates was unexpected, and led Rous to subject his techniques to rigorous tests to ensure that all cells and bacteria had been eliminated. Tumour number 3 was also a spindle-cell sarcoma, but histologically distinct from number 1 (Rous *et al.,* 1912). Number 7, an osteo-chondrosarcoma, produced true cartilage and bone, characters retained in secondary tumours produced by filtrates; this showed that the unusual features of this tumour were dependent on the properties of the virus. Another type of tumour was transmitted by filtered material by Fujinami and Inamoto (1930) at Kyoto in Japan. This was a myxosarcoma (the myxoma is a gelatinous form of tumour). This tumour, also, had initially been transmitted by serial transplants without success over transmission by filtrates; such was, however, eventually achieved (Fujinami and Suzue, 1928; Fujinami, 1930). This tumour grew readily in ducks, which were killed by it; it could be transplanted back from the ducks into chickens.

Researches on the Rous Sarcoma Virus have continued to the present time. Particular breeds of chickens have been favoured for this work, because of their genetic susceptibility. In the United States, the New Hampshire red breed has been especially favoured, whereas in Britain the Edinburgh strain of the brown leghorn has been very successful. In both, younger birds give better results than older. Even within a single breed, as reported by Waters and Fontes (1960), there exist significant familial variations and susceptibility, which recalls the familial incidence of some cancers in human families. In relation to the human problem, it is also important to determine whether *all* chicken sarcomas are caused by viruses. If some of these tumours could be shown to have a non-infectious cause, it would give greater credence to the possibility that human sarcomas were also non-infectious. This point is discussed by Gross (1970, p. 112 *et seq.*). The evidence suggests that all tumours of this group do have an infectious origin, though sometimes there are difficulties in demonstrating it and in isolating the virus; often this can only be done, when the tumour has been passed serially in the first place by means of grafts. For example, Nebenzahl (1934) in a review of thirty spontaneous chicken tumours from various parts of the world concluded that all sarcomas of the fowl, which could be transplanted for a sufficient number of passages, would eventually yield a transmissible agent that could be recovered by filtration; it is not, however, always possible to reproduce the tumour by transplant. This view was supported by Murphy and Sturm (1941). There is also evidence that tumours induced by chemicals yield infectious virus, such as adduced by Carrel (1925) and McIntosh (1933, 1935). Other authors, such as Peacock (1933) were unable to confirm these results. The difficulties are highlighted by the experience of Oberling and Guérin (1950). These French authors described the appearance of a sarcoma in the foot of a hen 32 months after injection of an oily solution of methylcholanthrene. The tumour grew very slowly and could initially be transmitted only by grafts. At the fifth passage, 20 months after the appearance of the primary tumour, and thereafter, it became possible to transmit the tumour by filtered material.

The difficulties of obtaining infectious material from all tumours has been shown to rest with the concentration of viral particles in the tumour tissue by a number of workers, *vide* Bryan (1955, 1956); Bryan *et al.* (1954, 1955);

Moloney (1956); Riley *et al.* (1946). The concentration of virus in spontaneous tumours is invariably low, but increases with transplantation; sometimes, too, tumours that have shown filterable properties lose them for months on end, but recover them later. More virus, also, is present in younger tumours than older and in tumours from younger animals than older. It was found also by Bryan (1955) and Prince (1959) that tumours induced by very small doses of virus were liable to be non-infectious and Prince suggested that this might be because the virus was incomplete or defective, and unable to reproduce itself in the absence of a 'helper' virus. That this was so was demonstrated by Rubin and Vogt (1962) and Hanafusa *et al.* (1963); the helper virus was designated Rous Associated Virus (RAV). When chicken embryo cells are infected in tissue culture with Rous Sarcoma Virus (RSV) alone, they become transformed into Rous sarcoma cells but do not produce infectious particles; when RAV is added, infectious particles are produced.

Chicken leukaemia viruses can themselves at times act as RAV; at other times they appear to inhibit the development of sarcoma by these viruses. As with the sarcoma viruses, their virology has been extensively studied by advanced and ingenious techniques and a great deal is coming to be known about them and other oncornaviruses. There is strong evidence that all the chicken leukaemias and sarcomas result from infection with viral agents. This cannot unfortunately be categorically stated, because virus cannot always be isolated. In relation to human cancer, the most important observations are likely to be concerned with the infectivity of these viruses for non-human primates. Munroe and Windle (1963) injected 4 adult and 7 newborn rhesus monkeys (*Macaca mulatta*) with minced suspensions of Rous Sarcoma. All the newborn, but none of the adults, developed fibro-sarcomas. Zilber *et al.* (1965) injected cell suspensions of the Carr–Zilber strain of RSV into 4 newborn monkeys (2 rhesus, 1 pigtail (*M. nemestrina*), and 1 baboon (*Papio hamadryas*). All of these monkeys developed fibro-sarcomas; the tumours regressed in all the monkeys, except in pigtail, which developed a very large tumour and died. Deinhardt (1966, 1969) showed that both young and adult marmosets (*Saguinus* spp.) develop rapidly growing fatal sarcomas from RSV infection, which becomes metastasised in lymph nodes, lungs, kidneys and brain, see chapter 7.

III The leukosis complex in chickens

The leukosis complex, also caused by oncornaviruses, is in some ways also associated with sarcomas and renal carcinomas. Ellerman and Bang (1908, 1909) recorded that some of the chickens injected with filtered material developed typical generalised leukaemia; others developed leukaemic changes in internal organs, infiltration with leukaemic cells, swelling of spleen and liver and other changes characteristic of leukaemia, except that the blood did not show leukaemic changes; this form of the disease, they termed 'pseudo-leukaemia' or 'alaeukaemic leukaemia'. It was they, who coined the term 'leukosis' to cover both forms of the disease. The same filtrates would in one case cause the leukaemic disease and in others the aleukaemic. Furthermore, sometimes injection of leukaemic filtrates resulted in the development of sarcomas, and grafts of such sarcomas would again cause leukaemias; sometimes, indeed, leukaemia and sarcoma might come to co-exist as a result of their endeavours. These aspects of the problem were studied further by Oberling and Guérin (1933, 1934), Engelbreth-Holm (1942), Furth (1931, 1933, 1936) and Furth and Stubbs (1934). Carr (1956) found that an erythroblastic leukaemia strain of virus induced multiple adenocarcinomas in brown leghorn chicks, if they were given infectious material when under two weeks old. The problem to be considered was, thus, whether these were varying manifestations of the activity of a single virus, which seemed unlikely, whether a single virus was involved which readily changed its character by mutation, or whether there existed a complex of related viruses which might be present singly in the body or co-exist. As with RSV, there are breed susceptibilities to the virus or viruses; the Red New Hampshire, used so successfully with RSV studies, is very resistant, but white leghorns are susceptible. With these birds, it was neatly and satisfactorily shown that erythroblastosis and myeloblastosis are caused by distinct, but related, viruses. That of myeloblastosis possesses a readily detectable enzyme by which adenosinetriphosphate (ATP), a host enzyme important in cellular energy exchanges, is dephosphorylated, that is deprived of its phosphorus molecule; the virus of erythroblastosis contains no such enzyme (Bonar *et al.*, 1957). Again, 3-day-old chicks are most susceptible to the myeloblastosis virus and resistance is increasingly enhanced with age. With erythroblastosis, the reverse is true, since susceptibility progressively increases from the seventh day onwards, being maximal at 2–3 weeks. Antigenically, on the other hand, the two viruses are closely related and protection against infection can be given with immune sera from either (Eckert *et al.*, 1955; Beard, 1957; Beard *et al.*, 1957).

The identity of the agent, which caused visceral lymphomatosis, was established by Burmester *et al.* (1946). They, too, faced difficulties, since it appeared initially that different related and unrelated viruses were present together in some filtrates. For example, of 189 chicks that were injected with filtered material 81% developed tumours within 6 months; of these 88% had visceral tumours, but 55% also developed osteopetrosis and 6% neurolymphomatosis (Marek's Disease). Filtrates from donors with osteopetrosis only induced both visceral tumours and osteopetrosis; some of those with visceral tumours only produced osteopetrosis also. The virus – or viruses – most readily infect 1–2-day-old chicks, but positive results can be obtained in much older birds up to 114 days. Visceral lymphomatosis can be horizontally transmitted from birds showing symptoms of infection, or from 'carrier' birds which show no symptoms; the virus can be isolated from the saliva, but carrier birds can also transmit the virus vertically in the eggs. Virus is found in greatest concentration in the liver and kidneys, but particles are also present in large numbers in the female reproductive system and in lesser concentration in spleen, intestines, muscle, brain and blood. On the evidence available so far, it appears that the virus which causes visceral lymphomatosis causes also osteopetrosis and kidney tumours and this may be the same as the virus of erythroblastosis. If different viruses are involved, they are evidently closely related, and the probability is that gene exchange between them is of common occurrence. The relationship of the leukaemia viruses with RSV adds to the complexity of the picture. It is certain, however, that there exist in chickens two families of related neoplastic diseases, which are undoubtedly caused by viruses, the sarcomas and some carcinomas, and the leukaemic diseases of the blood and blood-forming tissues. These are counterparts of similar diseases in other groups of animals, including man, and these too require to be studied to discover whether they also have infectious origins. We must first, however, study those neoplastic conditions in chickens, which are caused by other viral groups.

IV Marek's Disease

Neural lymphomatosis (fowl paralysis or Marek's Disease) was first described by Marek (1907). In affected birds, the nerve trunks become thick and swollen because of infiltration with lymphoid cells. In the early stages, the wings droop and there is weakness and lack of co-ordination in the legs. The birds then become unable to stand and complete paralysis ensues. The disease is endemic in some areas and may become epidemic causing serious losses. Younger birds are more frequently affected than older. The causative virus cannot be demonstrated directly in tumours or in the blood, and must be isolated in tissue culture (Churchill and Biggs, 1967; Soloman *et al.*, 1968; Nazerian, 1973). The virus belongs to the herpesvirus group, many of which are associated with cancers in animals including man. It was shown by Calnek *et al.* (1970) and Nazerian and Witter (1970) that the true habitat of the virus is in the epithelial cells of the feather follicles. It is only in this site that the virus can come to maturity and reproduce itself. Its location in the nerve trunks is aberrant although not unexpected, because many herpesviruses, including that which causes chicken pox in man, also become latent in nerve tissues. Churchill *et al.* (1969) described the immunisation of day-old chicks, so as to give lifelong protection to the birds. There are in use today four strains of vaccine, as described by Nazerian (1973). The most effective is an antigenically related turkey strain of the virus (Kawamura *et al.*, 1969; Witter *et al.*, 1970). Vaccinated birds do not resist infection with the oncogenic virus, but become infected as if they had not been immunised. Both viruses, indeed, persist in the body for life, being continuously shed in feather follicle epithelium. In some manner unknown, the heterologous virus inhibits the cancer-causing potential of the chicken virus.

Marek's Disease is of profound importance, since not only is this a cancer unquestionably caused by a virus of the herpes group, but also the disease is the only cancer so far to be efficiently controlled by a vaccine in commercial production. All of the chicken cancers are of profound importance as providing a more or less complete model for similar diseases in animals and man, though a model for mammary tumours had to await the elucidation of similar problems in mice. I quote *verbatim* from Maurice R. Hilleman (1974), the well-known American expert on the cancers and a proponent of the view that vaccines will in time be developed against them:

> It may be worthy of note that the foundations for modern knowledge of mammalian RNA leukaemia and sarcoma viruses are based on the pioneering studies carried out with chicken viruses in the chicken model system. It seems true also that the further progress with oncogenic mammalian herpesviruses will have its fountain in the advances that have been made with Marek's Disease in chickens. The domestic fowl has provided fine models for studies of cancer and the contributions that have been made are something worth 'crowing' about.

Before passing to the viral causes of mammalian cancers, we shall study an important cancer of frogs, the Lucké Renal Adenocarcinoma, also caused by a virus of the herpes group.

V The Lucké Tumour of frogs

The frog renal adenocarcinoma (Lucké Tumour) was first described by Bernard Lucké (1934, 1938). The tumours occur in wild leopard frogs (*Rana pipiens*), of which some 2.7% of 10 000 frogs were found to be affected in the northern New England states, especially the Lake Champlain region of northern Vermont and bordering areas of Quebec province in Canada. Affected frogs have also been found in North Dakota, Indiana and the Mississippi valley. Lucké considered that the tumours must be caused by a virus, because, when examined histologically, 'inclusion' bodies characteristic of viral infections were found in the nuclei of many of the tumour cells. Subsequently, Fawcett (1956) described typical herpesvirus particles in many of the tumour cells, and Lunger (1964) isolated the causal virus.

The Lucké Tumour is important, because it was the first tumour to be associated with a virus of the herpes group, and in this way gave to lead to investigators of tumours in higher animals, such as the Epstein–Barr Virus (EBV) in man (Epstein, 1962). It is also important, because of the behaviour of the virus in a cold-blooded animal, which might lead to a better understanding of the properties of these viruses. Tumours in frogs caught in the summer and maintained in the laboratory at a temperature of 20–25°C are designated 'summer' tumours; those in frogs caught in the winter or maintained at temperatures of 4°C–9°C are designated 'winter' tumours. The summer tumours contain no virus inclusion bodies or demonstrable viral particles; the winter tumours begin to degenerate, exhibit inclusion bodies and release infectious virus. Experiments reported by Rafferty (1965), Mizell *et al.* (1968) and Stackpole (1969) have shown that, when summer tumour tissue is maintained at low temperatures, virus production is started. Thus the Lucké virus genome becomes active at low temperatures, when the host genetic material is relatively inactive during hibernation. *Herpesvirus simplex,* the cause of herpes in man during childhood, has also a temperature sensitivity, since it causes the skin lesions, known as 'cold sores', during febrile periods. In this respect, there is a similarity with the frog virus.

VI Conclusion

The other groups of carcinogenic viruses which occur in chickens may be ignored, since in them they are unimportant and we shall encounter them in other species. After more than forty years, it was found that the malignant tumours of mice exhibited an identical pattern with that of chickens. It had been held by many that, if so wide a spectrum of avian cancers was demonstrably due to infectious causes, the same must be true of neoplastic diseases of mammals. Such beliefs have been vindicated.

Today, it is evident that most, if not all, mammalian tumours are caused by viral agents, with the sole exception of man. The long haul from chicken to mouse and then to other mammals is repeated in the mammal to man trail. In spite of all the odds against, scepticism prevails as to whether man is not the exception which proves the rule, and that his cancers are due to other causes. However, man is not available as an experimental animal and, as we have seen, a prerequisite for success with chickens was the serial transplantation of tumour material from chicken to chicken before viruses could be isolated from filtered material. Success with mice, as we shall see in the next chapter, was dependent on the production by inbreeding of races of mice that were especially susceptible to the cancer that was being studied. Nevertheless, important advances have been made, and some at least of the human cancers are generally believed to be of viral origin in the light of strong evidence that this is so. Knowledge of the viruses responsible is essential to devising means of protection against them, as has been achieved with Marek's Disease of chickens.

In the next chapter we shall follow the history of research into the neoplasias of mice and other rodents, and in succeeding chapters see how far these systems are applicable to other animals.

References

Beard, J. W. (1957). Etiology of avian leukosis. *Ann. NY Acad. Sci.* **68**, 473–486

Beard, D. G., Beaudreau, G. S., Bonar, R. A., Sharp, D. and Beard, J. W. (1957). Virus of avian erythroblastosis. III Antigenic constitution and relation to the agent of avian myeloblastosis. *J. nat. Cancer Inst.* **18**, 231–259

Bittner, J. J. (1936). Some possible effects of nursing on the mammary gland tumour incidence in mice. *Science* **84**, 162

Bonar, R. A., Beaudreau, G. S., Sharp, D. G., Beard, D. and Beard, J. W. (1957). Virus of avian erythroblastosis. V Adenosine-triphosphatase activity of blood plasma from chickens with the disease. *J. nat. Cancer Inst.* **19**, 909–922

Bryan, W. R. (1955). Biological studies on the Rous Sarcoma Virus. I General introduction. II Review of sources of experimental variation and methods for their control. *J. nat. Cancer Inst.* **16**, 285–315

Bryan, W. R. (1956). Biological studies of the Rous Sarcoma Virus. IV Interpretation of tumour response data involving one inoculation site per chicken. *J. nat. Cancer Inst.* **16**, 843–863

Bryan, W. R., Moloney, J. B. and Calnan, D. (1954). Stable standard preparations of the Rous Sarcoma Virus preserved by freezing and stroring at low temperatures. *J. nat. Cancer Inst.* **16**, 317–335

Bryan, W. R., Calnan, D. and Moloney, J. B. (1955). Biological studies on the Rous Sarcoma Virus. III The recovery of virus from experimental tumours in relation to initiating dose. *J. nat. Cancer Inst.* **16**, 317–335

Burmester, B. R., Pricket, C. O. and Belding, T. C. (1946). A filterable agent producing lymphoid tumours and osteopetrosis in chickens. *Cancer Res.* **6**, 189–196

Calnek, B. W., Adldinger, H. K. and Kahn, D. E. (1970). Feather follicle epithelium: a source of enveloped and infectious cell-free herpesvirus from Marek's Disease. *Avian Dis.* **14**, 219–233

Carr, J. G. (1956). Renal adenocarcinoma induced by Fowl Leukaemia Virus. *Brit. J. Cancer* **19**, 379–383

Carrel, A. (1925). Un sarcome fusocellulaire produit par l'indol et transmissible par un agent filtrant. *Compt. rend. Soc. Biol.* **93**, 1278–1280

Churchill, A. E. and Biggs, P. M. (1962). Agent of Marek's disease in tissue culture. *Nature* **215**, 528–530

Churchill, A. E., Payne, L. N. and Chubb, R. C. (1969). Immunization against Marek's Disease using a live attenuated virus. *Nature* **221**, 528–530

Deinhardt, F. (1966). Neoplasms induced by Rous Sarcoma Virus in New World monkeys. *Nature* **210**, 443

Deinhardt, F. (1969). Current trends in use of marmosets in virological research. *Ann. NY Acad. Sci.* **162**, 551–555

Eckert, E. A., Sharp, D.G., Beard, D., Green, I. and Beard, J. W. (1955). Virus of erythroblastic leukosis. III Antigenic constitution and immunologic characterization. *J. nat. Cancer Inst.* **16**, 593–653

Eddy, B. E., Borman, G. S., Grubbs, G. E. and Young, R. D. (1962). Identification of the oncogenic substance in rhesus monkey kidney cell cultures as Simian Virus 40. *Virology* **17**, 65–75

Eiselsberg, A. von (1890). Über einen Fall von erfolgreicher Transplantation eines Fibrosarkoms bei Ratten. *Wiener klin. Wochenschr*, **3**, 927–928

Ellerman V. and Bang, O. (1908). Experimentelle Leukämie bei Hühnern. *Centralb. f. Bakt. Abt. I* (Orig). **46**, 595–609

Ellerman, V. and Bang, O. (1909). Experimentelle Leukämie bei Hühnern. *Zeitschr. f. Hyg. v. Infektionskr.* **63**, 231–272

Engelbreth-Holm, J. (1942). 'Spontaneous and Experimental Leukaemia in Animals'. 245 pp. Edinburgh: Oliver & Boyd

Epstein, M. A. (1962). Observations on the mode of release of herpes virus from infected HeLa cells. *J. Cell Biol.* **12**, 589–597

Fawcett, D. W. (1956). Electron microscope observations on intracellular virus-like particles associated with the cells of the Lucké renal adeno-carcinoma. *J. Biophys. Biochem.* **2**, 725–742

Finkel, M. P., Birkis, B. O. and Jinkins, P. B. (1966). Virus induction of osteosarcomas in mice. *Science* **151**, 698–700

Firket, C. (1892). De la réussite de greffes sarcomateuses en série. *Bull. de l'Acad. Royale de Med. de Belgique* (IV series) **6**, 147–148

Fujinami, A. (1930). Special report: a pathological study in chicken sarcoma. *Trans. Japan Path. Soc.* **20**, 3–38

Fujinami, A. and Inamoto, K. (1930). Über Geschwülste bei japanischen Haushühnern insbesondere über einen transplantablen Tumor. *Zeitschr. f. Krebsforsch.* **14**, 94–119

Fujinami, A. and Suzue, K. (1928). Contribution to the pathology of tumour growth. Experiments on the growth of chicken sarcoma in the case of heterotransplantation. *Trans. Japan Path. Soc.* **15**, 281–282.

Furth, J. (1931). Erythroleukosis and the anemias of the fowl. *Arch. Path.* **12**, 1–30

Furth, J. (1933). Lymphomatosis, myelomatosis, and endothelioma of chickens caused by a filterable agent. I Transmission experiments. *J. exp. Med.* **58**, 253–275

Furth, J. (1936). The relation of leukosis to sarcoma of chickens. II Mixed osteochondrosarcoma and lymphomatosis (strain 12). III Sarcomata of strains 11 and 15 and their relation to leukosis. *J. exp. Med* **63**, 127–155

Furth, J. and Stubbs, E. L. (1934). Tissue culture studies on relation of sarcoma to leukosis of chickens. *Proc. Soc. exp. Biol. Med.* **32**, 381–383

Girardi, A. J., Sweet, B. H., Slotnick, V. B. and Hilleman, M. R. (1962). Development of tumours in hamsters inoculated in the neonatal period with vacuolating virus, SV 40. *Proc. Soc. exp. Biol. Med.* **109**, 649–660

Gross, L. (1951). 'Spontaneous' leukaemia developing in C_3H mice following inoculation in infancy with Ak-leukaemic extracts or Ak embryos. *Proc. Soc. exp. Biol. Med.* **78**, 27–32

Gross, L. (1953). A filterable agent, recovered from Ak leukaemic extracts, causing salivary carcinomas in C_3H mice. *Proc. Soc. exp. Biol. Med.* **83**, 414–421

Gross, L. (1970). 'Oncogenic Viruses', 2nd edn, 991 pp. Oxford: Pergamon Press

Hanafusa, H., Hanafusa, T. and Rubin, H. (1963). The defectiveness of Rous sarcoma virus. *Proc. nat. Acad. Sci. USA* **49**, 572–580

Hanau, A. (1889). Erfolgreiche experimentelle Übertragung von Carcinom. *Fortschr. d. Med.* **7**, 321–339

Hilleman M. R. (1974). Prospects for vaccines against cancer. *In* (Kurstak, E. and Maramorosch, K., eds) 'Viruses, Evolution and Cancer'. New York and London: Academic Press

Iwanowski, D. (1894). Über die Mosaikkrankheit der Tabakspflanze. *Bull. Acad. imp. Sci., St. Petersburg,* n.s. 3(**35**), 67–70

Iwanowski, D. (1899). Über die Mosaikkrankheit der Tabakspflanze. *Centralb. f. Bakt., Abt. II* **5**, 250–254

Jarrett, W. F. H., Martin, W. B., Crighton, G. W., Dalton, R. G. and Stewart, M. F. (1964). Leukaemia in the cat. Transmission experiments with leukaemia (lymphosarcoma). *Nature* **202,** 566–567

Kawamura, H., King, D. J. and Anderson, D. P. (1969). A herpesvirus isolated form kidney cell culture of normal turkey. *Avian Dis.* **13**, 853–863

Lucké, B. (1934). A neoplastic disease of the kidney of the frog, *Rana pipiens. Am. J. Cancer* **20**; 352–379

Lucké, B. (1938). Carcinoma of the kidney in the leopard frog: the occurrence and significance of metastasis. *Am. J. Cancer* **341**, 15–30

Lunger, P. A. (1964). The isolation and morphology of the Lucké frog kidney tumour virus. *Virology* **24**, 138–145

Marek, J. (1907). Multiple Nervenentzündung (Polyneuritis) bei Hühnern. *Deutsche Tierärztl. Wochenschr.* **15**, 417–421

McIntosh, J. (1933). On the nature of the tumours induced in fowls by injections of tar. *Brit. J. exp. Path.* **14**, 422–434

McIntosh, J. (1935). The sedimentation of the virus of Rous sarcoma and the bacteriophage by a high-speed centrifuge. *J. Path. Bact.* **41**, 215–217

Mizell, M., Stackpole, C. W. and Halperen, S. (1968). Herpes-type virus recovery from 'virus-free' frog kidney tumours. *Proc. Soc. exp. Biol. Med.* **127**, 808–814

Moloney, J. B. (1956). Biological studies on the Rous sarcoma virus. V Preparation of improved standard lots of the virus for use in quantitative investigations. *J. nat. Cancer Inst.* **16**, 877–878

Morau, H. (1891). Inoculation en série d'une tumeur épitheliale de la souris blanche. *CR. Soc. Biol.* **43**, 289–290

Munroe, J. S. and Windle, W. F. (1963). Tumours induced in primates by chicken sarcoma virus. *Science* **140**, 1415–1416

Murphy, J. B. and Sturm, E. (1941). Further investigation of induced tumours in fowls. *Cancer Res.* **1**, 477–483

Nazerian, K. (1973). Marek's Disease: a neoplastic disease of chickens caused by a herpesvirus. *Adv. Cancer Res.* **17**; 279–315

Nazerian, K. and Witter, R. L. (1970). Cell free transmission and *in vivo* replication of Marek's Disease virus. *J. Virol.* **5**, 388–397

Nebenzahl, H. (1934) 'Étude expérimentale des tumeurs de la poule.' 165 pp. Thèse Librairie E. le François, Paris

Novinsky, M. (1876). Zur Frage über die Impfung der krebsigen Geschwülste. *Gentralb. f. med. Wissensch.* **14**, 790–791

Oberling, G. and Guérin, M. (1933). Nouvelles recherches sur la production des tumeurs malignes avec le virus de la leucémie transmissible des poules. *Bull. du Cancer.* **22**, 326–360

Oberling, C. and Guérin, M. (1934). La leucémie érythroblastique ou érythroblastose transmissible des poules. *Bull. du Cancer* **23**, 38–81

Oberling, C., and Guérin, M. (1950). Sarcome de la poule par méthyl-cholantrène devenu filtrable. *Bull. du Cancer.* **37**, 5–14

Peacock, P. R. (1933). Production of tumours in the fowl by carcinogenic agents. *J. Path. Bact.* **36**, 141–152

Prince, A.M.C. (1959). Quantitative studies on Rous Sarcoma Virus. IV An investigation of 'non-infective' tumours induced by low doses of virus. *J. nat. Cancer Inst.* **23**, 1361–1381

Rafferty, K. A. jr. (1965). The cultivation of inclusion-associated viruses from Lucké tumour frogs. *Ann. NY Acad. Sci.* **126**, 3–21

Riley, V. T., Calnan, D. and Bryan, W. R. (1946). Studies on the influence of age on the latent period response of chickens to the agent of chicken tumour I. *J. nat. Cancer Inst* **7**, 93–98

Rous, P. (1911). A sarcoma of the fowl transmissible by an agent separable from tumour cells. *J. exp. Med.* **13**, 397–411

Rous, P. and Murphy, J. B. (1913). Variations in a chicken sarcoma caused by a filterable agent. *J. exp. Med.* **17**, 219–231

Rous, P. and Murphy, J. B. (1914). On the causation by filterable agents of three distinct chicken tumours. *J. exp. Med.* **19**, 52–69

Rous, P., Murphy, J. B. and Tytler, W. H. (1912). The relation between a chicken sarcoma's behaviour and the growth's filterable cause. *J. Am. med. Assoc.* **58**, 1840–1841

Rubin, H. and Vogt, P. K. (1962). An avian leukosis virus associated with stocks of Rous Sarcoma Virus. *Virology* **17**, 184–194

Sanarelli, G. (1898). Der myxomatogene Virus. Beitrag zum Stadium der Krankheitserreger ausserhalb des sichtharren. *Centralb. f. Bakt. Abt. I* **23**, 865–873

Shope, R. E. (1932). A filterable virus causing tumour-like condition in rabbits and its relation to virus myxomatosum. *J. exp. Med.* **56**, 803–822

Shope, R. E. (1933). Infectious papillomatosis of rabbits. *J. exp. Med.* **58**, 607–624

Soloman, J. S., Witter, R. L., Nazerian, K. and Burmester B. R. (1968). Studies on the etiology of Marek's disease. I Propagation of the agent in cell culture. *Proc. Soc. exp. Biol. Med.* **127**, 173–177

Stackpole, C. W. (1969). Herpes type virus of the frog renal adenocarcinoma. I Virus development in tumor transplants maintained at low temperature. *J. Virol.* **4**, 75–93

Stewart, S. E., Eddy, B. E., Gochenour, A. M., Borghese, N. G. and Grubbs, G. E. (1957). The induction of neoplasms with a substance released from mouse tumours by tissue culture. *Virology* **3**, 380–400

Sweet, B. H. and Hilleman, M. R. (1960). The vacuolating virus SV40. *Proc. Soc. exp. Biol. NY* **105**, 420

Trentin, J. J., Yabe, Y. and Taylor, G. (1962). Tumour induction in hamsters by human adenovirus. (abstract) *Proc. Am. Assoc. Cancer Res.* **3**, 369

Velich, A. (1898). Beitrag, zur Frage nach der Übertragbarkeit des Sarcomes. *Wien. med. Blätter* **21**, 711–712; **21**, 729–731

Waters, N. F. and Fontes, A. K. (1960). Genetic response of inbred lines of chickens to Rous Sarcoma Virus. *J. nat. Cancer Inst.* **25**, 351–357

Wehr, W. (1888). Demonstration der durch Impfung von Hund auf Hund erzeugten Carcinomknötchen. *Centralb. f. Chir. Suppl. 24,* **15**, 8–9

Witter, R. L., Nazerian, K., Purchase, H. G. and Burgoyne, G. H. (1970). Isolation from turkeys of a cell-associated herpes-virus antigenically related to Marek's Disease Virus. *Am. J. vet. Res.* **31**, 525–538

Zilber, L. A., Lapin, B. A. and Adzighitov, F. I. (1965). Pathogenicity of Rous Sarcoma Virus for monkeys. *Nature* **205**, 1123–1124

5

The Transmissible Tumours of Mice and Rabbit Papilloma

I The murine leukaemias

The transmission of leukaemia in mice by means of cell-free filtrates was first reported by Ludwik Gross (1951a,b). Unlike chickens, there were initially no separate breeds and races of mice, some very susceptible to the leukaemias and others resistant. Success in leukaemia research requires: first, a susceptible breed to provide an adequate supply of leukaemic material; secondly, a breed with low tumour incidence, in which to test this material. Leukaemic tumours, though never completely absent from wild and unselected mice, were too scarce to provide the material required, and the concentration of virus in naturally occurring tumours is too low for success. It took some twenty years for inbred strains of mice to be developed. Those most used in cancer research are shown in Table 1.

By the time Gross embarked on his experiments in 1946, there were suitable mice available for experiment with both high and low leukaemia

Table 1 Some of the inbred lines of mice more frequently used in cancer research. (After Gross, 1970)

Symbol	Colour	Mammary tumour % incidence	Leukaemia % incidence	Reference
A	Albino	90	1	Strong (1936)
Ak	Albino	1	85	Furth *et al.* (1933)
BALB	Albino	2	15	Bagg (1925)
CBA	Black agouti	3	1	Strong (1942)
C3H	Black agouti	90	0.5–1	Strong (1935)
C57	Brown	4	5	Little (1947)
C57	Black	1	5	Little (1947)
C58	Black	1	85	McDowell & Richter (1935)
DBA2	Dilute brown	60	30	Little (1958)
I	Dilute brown	1	5	Strong (1942)
RIII	Albino	75	3	D-Zavadskaia (1933)
Swiss	Albino	15–20	5–11	Not inbred or only partially

incidence. In the meanwhile work had proceeded on the induction of leukaemia in mice by chemicals, hormones and radiation. This could readily be achieved even in mice of low susceptibility, most readily by means of the known carcinogenic (cancer-inducing) hydrocarbons, such as dibenzanthracene (Burrows and Cook, 1936), or benzpyrene (Furth and Furth, 1938). Lacassagne (1937) reported that prolonged treatment of low-leukaemic mice with oestrogenic hormones would induce leukaemia. This work was successfully repeated by Gardner (1937) and Gardner *et al.* (1940, 1944); up to 12% of the mice developed leukaemia against a natural incidence of 0–5%. Radiation enhanced the action both of the chemicals and the oestrogens (Kirschbaum *et al.*, 1953; Kawamoto *et al.* 1958).

Leukaemia had also been transmitted in inbred strains by cell grafts, as described by Richter and McDowell (1929). This was achieved in C58 mice, but other mouse strains were unsuccessful (Richter and McDowell, 1929, 1930). Successful results were also achieved by Furth and Strumia (1931) using the A strain, from which the Ak was later developed; these transplants were successful in both related and unrelated families of mice.

In spite of these successes, numerous attempts to transmit leukaemia by means of filtrates had ended in failure, even when radiation was administered as an adjunct. These negative results had led to the virtual abandonment of the idea that these leukaemias had a viral cause, in spite of their homology with the chicken leukaemias. By 1949, Gross himself was ready to abandon his experiments after three frustrating years. As he subsequently realised, there were good reasons for his failures. These were: 1. the mice he used for his experiments at 7–10 days old were too mature and suckling mice should have been used, 2. his C3H mice were of a relatively insusceptible line, 3. he used Seitz asbestos filter pads, which absorb particles of some viruses, so that they do not appear in the filtrates. Earthenware or porcelain filters would have been better, and 4. his leukaemic extracts were prepared from single mice instead of pooled tissues from a number.

Gross then started to use 12-hour-old C3H mice, but abandoned for the time being his filtration procedures. After coarse filtration of the finely ground organs, he centrifuged the material at 3000 r.p.m. for 15 minutes. Using the supernatant fluid from such preparations, he induced a 46% leukaemia incidence in his experimental animals, which developed in 3½–11½ months; the expected incidence was 0.6%–1.0%. Unfortunately with this technique there could be no guarantee that some transformed cells had not been included in the injection. In 1951, he realised that he must revert to filtration, if he were to prove the presence of a virus. He then discarded his Seitz filters in favour of earthenware and porcelain (Berkefeld and Selas), using also material first centrifuged at 9500 r.p.m. for five minutes. With material prepared in this way, he achieved a high degree of success, up to

28%. A few of his mice also unexpectedly developed carcinomas of the parotid salivary glands (the glands that become swollen in mumps); this was due to a contaminant, the 'polyoma' virus (Gross, 1953), which will be considered later in this chapter. Gross had initially obtained his leukaemic material from affected mice of the highly susceptible Ak strain and was latterly using the Bittner subline of C3H mice as test animals. The leukaemia incidence was, nevertheless, relatively low, because as was later shown the concentration of viral particles in his material was low. Once the virus was well established by serial passage, the virus regularly caused leukaemia or lympho-sarcoma in most of the mouse strains tested.

Gross' original reports (1951a,b) were received at first with scepticism e.g. Furth (1951, 1952), and his techniques were criticised by Law (1954); indeed, Law *et al.* (1955) and Stewart (1955) attempted to repeat Gross' experiments at the National Cancer Institute, but without success. Further failures were reported by Nariaki Ida (1957) working at Dr Kirschbaum's laboratory at Baylor University in Houston and Kirschbaum (1957a,b) still stressed that multiple factors were no doubt responsible for the development of leukaemia, and the presence of a virus was not necessarily one of them. Confirmatory evidence was slow in coming, but success was eventually reported by Woolley and Small (1956) in New York, by Furth *et al.* (1956), Dulaney *et al.* (1957), and Hays and Beck (1958). Gross was, therefore, vindicated, and went on to develop a highly and constantly potent strain of leukaemia virus, termed virus A. Results with this virus are given in Table 2.

Table 2 Susceptibility of mice of different strains to leukemogenic action of passage A filtrates. (After Gross, 1970)

Mouse strain	Serial pass. no.	No. injected	No. +	% incidence	Age mths
C3H/Bi	27–28	189	187	99	2.7
C3H/An	21	37	26	70	3.9
C57/BR/cd	25–28	40	38	95	2.8
BALB/c	26–28	41	36	88	4.3
Swiss	30–31	53	42	79	3.5
I	29–30	17	11	65	5.2

A number of different forms of leukaemia developed in mice injected with leukaemia virus A, as had happened with the chicken leukaemia virus. Gross (1970) gives the proposed classification shown in Table 3.

Since the discoveries by Gross became accepted, the isolation of murine leukaemia viruses (MuLV) has become fashionable. As expressed by Bentzvelen (1974):

Table 3 Forms of leukaemias and lymphomas induced in mice and rats with mouse leukaemia virus. (After Gross, 1970)

	LEUKAEMIAS
1	*Lymphatic leukaemia* (*i*) Aleukaemic (*ii*) Leukaemic
2	*Stem cell leukaemia*
3	*Myelogenous leukaemia* (*i*) Undifferentiated (*ii*) Well differentiated
4	*Chloro-leukaemia**
5	*Erythroblastic leukaemia**, atypical
6	*Monocytic-like leukaemia*
	LYMPHOMAS
1	*Lympho-sarcomas* (*i*) Generalised (*ii*) Local
2	*Reticulum-cell sarcomas*
3	*Hodgkin's-like lesions*

* These neoplastic conditions of the red blood cells and their precursors were only seen in mice, from which the thymus glands had been surgically removed.

> Every self respecting scientist in the mouse leukaemia field seems to have felt it necessary to have isolated his own leukaemia virus, which resulted in an infinite list of MuLV strains.

The most widely used strains are: 1. the Gross (GLV) – Gross (1951), 2. Friend virus (FLV), an erythroblastosis virus isolated by Charlotte Friend (1956, 1957), 3. Graffi virus (GiLV), a myeloid leukaemia virus isolated from a reticulum cell sarcoma by Graffi (1957), 4. Moloney Virus (MoLV), a lymphatic leukaemia virus from an anaplastic sarcoma isolated by Moloney (1959, 1960), 5. Radiation–induced lymphoma virus (RadLV), isolated by Lieberman and Kaplan (1959), and 6. Rauscher virus (RLV), an erythroblastosis virus isolated by Rauscher (1962).

Hartley *et al.* (1965) classified MuLV in three categories on the basis of growth characteristics in cell cultures: 1. the virus grows well in BALB/c mice cells, but not in NIH/Swiss, categorised as B-tropic, 2. grows well in NIH/Swiss mouse cells but not in BALB/c – N-tropic, and 3. grows well in both kinds of cells – NB-tropic. These observations are similar to those made by Kaplan (1967). On the basis of these observations, Table 4 classifies the MuLV viruses.

Tables 3 and 4 reveal a range of cancers caused by MuLV, fully comparable with those caused by chicken leukaemia viruses and of the same pattern.

When Gross started his work on the murine leukaemia viruses in 1945, he was a member of the US Army Medical Corps, assigned to the Veterans Administration Hospital in the Bronx, New York City. He had only his spare time to devote to the work, and could only use an unwanted room in the hospital basement. It is easy with hindsight to criticise the rather obvious defects in his early techniques, by which his success was delayed. Nevertheless, it was his conviction and persistence, which produced this important advance in cancer research by showing that a complex of mammalian cancers was of infectious origin. He did this at a time, when repeated failures had led to a virtual abandonment of research in this direction. Once the infectious origins of murine cancers had been accepted, the same was quite quickly shown to be true in cancers of other animals, because the direction of research was switched to the study of possible viral causes.

Table 4 Laboratory strains of murine leukaemia virus. (After Bentzvelen, 1974)

Name	Abbrev.	Predominant lesion	Host range	New cellular antigen
Gross leuk. v.	GLV	Lympho-sarcoma	N-tropic	G
Friend leuk. v.	FLV	Rythroblastosis	N-tropic	FMR
Graffi leuk. v.	GiLV	Myeloid leukaemia	NB-tropic	FMR
Moloney leuk. v.	MoLV	Lympho-sarcoma	NB-tropic	FMR
Rad. leuk. v.	RadLV	Lymphatic leukaemia	B-tropic	G
Rauscher leuk. v.	RLV	Erythroblastosis	NB-tropic	FMR

II The murine sarcoma viruses

The murine leukaemias, as with the chicken leukaemias, are caused by viruses of the oncornavirus group; similar viral particles can be demonstrated in high yielding tumour tissue and in productive tissue cultures. It remained to ascertain whether murine sarcomas, like the Rous and related tumours of chickens, were also caused by oncornaviruses. Again progress was slow, and it was not until 1964 that this was demonstrated to be so. It was subsequently shown also that, as with the chicken sarcoma viruses, those of mice are also 'defective', requiring the presence of leukaemia virus as 'helper'.

Dr Jennifer Harvey (1964) of the Department of Cancer Research at the London Hospital Medical School had injected rats with the Moloney strain of MuLV. Blood plasma from a rat, which developed leukaemia, was filtered and injected into newborn BALB/c mice and newborn rats. After only one month, both rats and mice developed pleomorphic sarcomas near the site of injection, but did not develop leukaemia. All of the animals, however, developed very large spleens, and some died of splenic rupture. Virus was readily recovered from tumours of rats and mice; hamsters also developed sarcomas, but virus could not be obtained from them. The virus could also be readily propagated in tissue culture and reproduced the disease when injected into newborn rats and mice; the mice, however, developed spenomegaly only without tumours (Harvey *et al.*, 1964; Chesterman *et al.*, 1966). Following Harvey's initial success, murine sarcoma viruses have been isolated by other workers. Moloney (1966a, b) injected large doses of Moloney strain mouse leukaemia virus into newborn BALB/c mice. As a result, the mice developed rhabdo-myosarcomas (muscle tumours) near the site of injection after a short latent period; this virus did not infect rats. A virus was isolated from lymphomas in old C3H mice, which caused erythroblastosis in mice after 25 days and lymphomas of the thymus in rats after 8–16 months. This same virus harvested from the rat lymphomas after further passage caused either erythroblastosis or disseminated lympho-sarcomas, when injected into newborn rats (Kirsten and Mayer, 1967; Kirsten *et al.,* 1968). When spleen extracts or plasma from rats with erythroblastosis were injected into newborn rats or C3H mice, they developed erythroblastosis *and* multiple pleomorphic sarcomas. Many of the rats also developed rarefying lesions of the skeleton, and both rats and mice with sarcomas had enlarged spleens suggestive of erythroblastosis. Gross (1963) and Gross *et al.* (1965) demonstrated further anomalies arising from the injection of the Gross A mouse leukaemia virus. During routine transmission for maintenance of the virus into newborn rats, myxomatous tumours developed instead of leukaemia; the affected animals also developed generalised lympho-sarcomas. In further experiments with this virus rhabdo-myosarcomas developed.

The mouse sarcoma viruses are evidently distinct from the mouse leukaemia viruses, though the two groups cannot be distinguished on the basis of physical and morphological characters, or sensitivity to heat and treatment with ether. Harvey (1969) expressed the opinion that sarcoma filtrates also contained leukaemia virus and other workers (Hartley and Rowe, 1966; Huebner *et al.,* 1966; Harvey and East, 1969) convinced themselves that the murine sarcoma viruses were defective, requiring the leukaemia virus as 'helper'.

Amongst the tumours of the chicken sarcoma complex was the disease of the skeleton known as osteopetrosis. In mice, too, there is a specific bone tumour condition, also caused by viral infection, the mouse osteo-sarcoma. Spontaneous bone tumours have a low incidence in wild or unselected mice, but became more numerous in certain inbred strains. Such were described by Pybus and Miller (1934) in laboratory mice of a strain known as the Simpson strain (Marsh, 1929). These mice also had a high incidence of mammary carcinomas and pulmonary and liver tumours. The tumour incidence was enhanced by further inbreeding and a 'special strain' was developed having a very high proportion of bone tumours; for example of 195 mice, which died after reaching the tumour age, 104 had bone tumours more commonly in females than males. The tumour developed in 15 months for females and 17 months for males. The tumours were frequently multiple and included osteomas, osteo-sarcomas, and spindle cell sarcomas (Pybus and Miller, 1938, 1940). In the light of newer knowledge, it is certain that all these tumours resulted from viral infections. However, at the time genetic causes of cancer were being sought and an infectious cause was not suspected. A search for such a virus was initiated by Finkel *et al.* (1966). In this search CF1 (Carnworth Farms line 1) mice were used, of which 1%–2% suffer naturally from bone tumours. The incidence is, however, augmented by administration of radioactive isotopes, such as strontium 90. The only virus recovered was from a spontaneous osteo-sarcoma in a 260–day-old male of the CF1 strain. Filtered material was injected into 5 newborn CF1 mice of the same litter, 2 of which developed osteo-sarcomas after 9 and 11 months. In the second passage 12 of 24 mice developed osteo-sarcomas within 9 months — 5 as early as 2½ months. In two years, 22 consecutive passages were made, from which only bone tumours developed. The tumours began to appear as early as 35 days, but some litters of mice were resistant. Examination of the tumours by electron microscope preparations revealed the presence of particles resembling those of the leukaemia and sarcoma viruses.

The various strains of mouse sarcoma virus can induce tumours in rats, as shown by Harvey (1964), Kirsten and Mayer (1967) and other authors, and also in hamsters (Harvey, 1964; Huebner *et al.* 1966, and other authors). In tissue cultures, these viruses grow in and transform rat cells (Ting, 1966), hamster cells (Bernard *et al.,* 1967; Simons *et al.,* 1967), and bovine cells (Thomas *et al.,* 1968). They can also transform human cells as shown by Bernard *et al.* (1969) and Aaronson and Todaro (1970). If the MSV is 'deficient' and, therefore, not producing infective particles, it can acquire the 'helper' from cells infected with leukaemia virus, not only of the same species but of other species also, such as cat cells. Aaronson (1971) and Bernard *et al.* (1972) have shown that non-productive MSV can acquire a 'helper' if co-cultivated with human tissue cells. This surprising fact argues strongly that the human cells must be harbouring a so far undiscovered leukaemia virus.

It is, therefore, fully established that mice in common with chickens suffer from complete complexes of leukaemic and sarcomatous diseases, cancers caused by distinct but related viruses of the oncornavirus group. Between them these animals have supplied model systems for the study of comparable diseases in other mammalian groups of animals, including man. Mice can, however, take us further along the path of knowledge, because they are mammals and possess mammary glands, the cancers of which might be expected to contribute to the study of human breast cancers, which are so important and cause such widespread trouble and distress.

III Mouse mammary carcinoma

If the attribution of the chicken leukaemias and sarcomas to viral infections is associated with the names of Ellerman and Bang, and of Peyton Rous, and that of the murine leukaemias with that of Ludwik Gross, the name associated with the mammary carcinomas of the mouse is that of John J. Bittner of the Roscoe B. Jackson Memorial Laboratory at Bar Harbor, Maine.

A very high incidence of mammary tumours had been observed in certain inbred strains of mice, especially the C3H strain, in which the incidence was 90% in both virgin and breeding females; the A strain of mice of Gross also showed an incidence of 90% in the breeding females, but only 5% in the virgins. The tumours were typical adenocarcinomas, and would appear at 5–12 months of age. In other strains, CBA, BALB/c, C57 Black, the incidence was very low. Very much earlier it had been observed e.g. by Lathrop and Loeb (1918) that the progeny of susceptible mothers and resistant fathers were as susceptible as the mothers; that this was so was confirmed in 1933 at the Jackson Memorial Hospital (*vide* Gross, 1970), and by Korteweg (1934, 1936) in Holland. Since no genetic factor for resistance was transmitted by the father, it was concluded that the cancer was transmitted by some factor not associated with the genetic apparatus of the mouse ('extra-chromosomal'). Such a factor might be passed to the offspring: 1. by way of the cytoplasm of the ovum, 2. during foetal life, or 3. by way of the mother's milk. The latter possibility was investigated by Bittner (1936, 1939, 1940a,b) and by Andervont (1941). The milk proved indeed to be the vehicle by which infection was conveyed and the agent responsible was termed the 'milk influence' or the 'milk factor'. It was found that newborn mice did not develop mammary carcinoma, if prevented from suckling their real mothers and transferred to foster mothers; conversely, newborn mice fostered on high incidence mothers developed breast cancers. This proved that mammary carcinoma of mice was an infectious disease transmitted by some agent present in the mother's milk. The low incidence in virgin mice of most strains and in males suggested that some hormonal factor was also involved. Such was proved to be the case by Murray (1928), who showed that, when male mice of a high tumour strain were castrated and had ovaries implanted in them, they duly developed mammary cancer. It was also found that some female mice, which did not develop cancer, nevertheless carried the virus throughout their lives and transmitted it to their offspring.

The infectious agent was isolated from filtered material by Bittner (1942) and by Andervont and Bryan (1944). Filtrates derived from the milk could cause the disease even in high dilutions, if given by mouth or injected intraperitoneally. On the other hand, only limited success was achieved over growing the virus in tissue cultures of mouse mammary cells by Lasfargues (1964), probably because essential glandular elements were lacking.

The mammary tumour agent is present both in the milk of nursing mothers and in tumour tissue. In addition, in mice of high tumour incidence the virus is widespread in the body organs. In this way, males can be carriers of the virus as well as females, to which they transmit it during coitus. This was proved by Andervont (1945), who mated low incidence BALB/c female mice with high incidence C3H males; of the offspring 60% of the females developed mammary carcinoma. Bittner (1952) repeated these experiments using the same mouse strains, reporting an incidence of 56% in the progeny and 96% in the inbred second generation. If the mammary carcinoma problem in mice has relevance to that of man, transmission of the disease by the father to the mother is of obvious importance. The disease would be venereal and promiscuous females would be at greater risk as with cervical carcinoma, a condition we shall study below, widely believed to be caused by a herpesvirus and also to be venereal.

Guérin (1955, 1956) examined mouse mammary tumour tissues for the presence of 'inclusion bodies' such as are found in the cells in many viral infections. He found them in 5 of 30 radiation induced tumours. In a further 52 mammary tumours of mixed origin, he found inclusions in 6. Finally, in 26 spontaneous mammary tumours in a high incidence strain of mice, there were 4 with inclusions. Porter and Thompson (1948) identified virus-like particles in mammary tumour cells by means of electron microscopy. They were found also in mouse milk, known by biological tests to contain the tumour agent, by Graffi *et al.* (1949) and Passey *et al.* (1950). Similar particles were found in the milk of women with a family history of breast cancer by Gross *et al.* (1950, 1952) and Dmochowski and Passey. (1952); similar particles were very rarely found in the milk of women without a familial history of breast cancer. Unfortunately, these accounts cannot be given full credence, because they were made from 'shadow-cast' electron microscope preparations, which does not permit precise details of viral morphology to be determined. The development of the ultra-microtome enabled ultra-thin sections to be prepared, whereby both cells and the viral particles could be precisely sectioned and positively identified. For an account of the identification of the virus, Gross (1970) should be consulted. Suffice it to say here, that positive identification of viral particles has been made in mammary tumour tissue, infected milk, and in the genital organs of male mice. Particles in human milk identical with those in mouse milk have also been identified. The virus was classified with the oncornavirus group, the particles being known as 'B particles', to distinguish them from 'A particles' also found in these preparations and regarded as an intermediate stage in the development of the mature virus. The B particles, although undoubtedly belonging to the oncornavirus group differ from those of mouse leukaemia and sarcoma, which are named 'C particles'. Kinosita *et al.* (1953) were the first to give a detailed description of the B particles in tumour tissue. They were followed by Dmochowski (1954) and Bernhard and Bauer (1955). In the male genital organs the viral particles were described by Moore (1963) and by Smith (1965). Young healthy mice of low incidence strains did not reveal the presence of viral particles in the mammary glands or any other tissues.

It may be observed that, in spite of what has been written above, Gross in his studies of mouse leukaemia was not the first to describe a viral cancer in a mammal. This honour lies with Bittner for his work on the 'milk factor' in mouse mammary carcinoma. Bittner's discovery, however, failed to achieve the impact of that of Gross. Students of cancer, Gross apart, were not stimulated to begin the search for viral agents in other cancers. The viral involvement in mammary carcinoma was regarded as but one of a number of factors including that of the sex hormones. Bittner did not, therefore, open the field to new perspectives, as did Gross. Earlier even than Bittner, Shope had involved viruses in the etiology of papillomas in rabbits, benign tumours which may become malignant. These will be studied at the end of this chapter. Meanwhile, we must turn our attention to an important tumour of mice, that of the parotid gland known as 'polyoma'. The causal agent belongs to the papovavirus group, so for the time being we are leaving the oncornaviruses.

IV Parotid gland tumour of mice (polyoma)

The papovavirus group is remarkable for the small size and weight of the viral particles. In many other respects, which will be studied later, they differ markedly from the oncornaviruses. In spite of this they originally appeared in murine leukaemia material, from which it was difficult to separate them. The members of this group, which have been shown to cause tumours, whether benign or malignant, are shown in Table 5.

Gross (1953) found that some Ak leukaemic filtrates, as stated above, injected into C3H mice caused tumours to develop on both sides of the neck instead of leukaemia. The tumours increased in size slowly but progressively,

Table 5 Polyoma and related viruses

Virus	Natural host
Polyoma virus	Mice
SV40 (simian virus 40)	Monkeys
Shope papilloma virus	Rabbits
Canine papilloma virus	Dogs
Bovine papilloma virus	Cattle
Human papilloma and wart virus	Man

until large collars enveloped the animals' necks. The tumours developed, when the animals were 3–4 months old, and consisted of small individual tumours up to the size of a pea; they were different from any other tumours hitherto described in mice. There were no signs of leukaemia and the tumours were located in the parotid salivary glands, unlike leukaemic tumours, and were of the nature of adenocarcinomas. At a later stage, affected mice often developed also subcutaneous tumours, which proved to be fibro-sarcomas or fibro-myxosarcomas: (Gross, 1953, 1955); some similar tumours also developed in the peritoneum, muscle or tissues surrounding the nerves, exhibiting peculiar features or being of the nature of rhabdo-myosarcomas.

Gross (1953) postulated that the parotid tumour virus was distinct from the leukaemia virus for the following reasons: 1. the parotid virus would pass through filters with a pore diameter small enough to retain leukaemia virus, 2. the leukaemia virus was precipitated in the ultra-centrifuge at much slower speeds than the parotid, showing it to be heavier, and 3. the leukaemia virus is destroyed at temperatures of 50°C–60°C in 30 minutes, whereas to destroy the parotid tumour virus 70°C for 30 minutes are necessary. Gross' findings were quickly confirmed by Law (1954) and Stewart (1953), who also induced parotid tumours in mice with leukaemic filtrates.

Gross (1955) attempted to pass parotid tumours by means of tumour tissue filtrates, but only 7% of the injected mice developed parotid tumours. Some of the mice developed leukaemia or subcutaneous fibro-sarcomas; a few developed both parotid tumours and fibro-sarcomas. Some few mice developed carcinomas in the submaxillary salivary glands and some in the medullary portion of the adrenal glands, neither of which had been found affected previously. Although positive results were few in number, they were statistically significant and justified the conclusion that the parotid tumour was caused by a viral agent.

The parotid gland virus was termed 'polyoma virus' by Stewart *et al.* (1957). Working at the National Institute of Health laboratories at Bethesda, they attempted to grow polyoma virus in cultures of rhesus monkey kidney cells, such as were routinely used in poliovirus vaccine production. In this medium, the polyoma virus grew well but the leukaemia virus was rejected, and pure uncontaminated virus became available for research. The virus was subsequently transferred to cultures of mouse embryo cells, which were better suited to it (Stewart *et al.*, 1958). In this medium, the virus not only grew with great vigour but increased in potency; this was quite unexpected, because the potency of no other virus had at that time improved in tissue culture. Typical of the results obtained with the culture virus were the following: of 67 Swiss mice injected when under 1 day old, 62 (92.5%) developed tumours at 3–4 months. All had tumours of the parotid gland; in addition some one third also suffered from submaxillary gland tumours and a few had sublingual salivary gland tumours also. Over 40% developed tumours of the thymus; 30% mammary carcinomas, and some 20% renal carcinomas; a few also developed sarcomas of the bones, carcinomas of the skin or digestive tract, adrenal medullary tumours, haemangiomas of the liver and others. Stewart *et al.* (1958) listed a total of 23 different tumours occurring in their mice. In all, however, the parotid tumour was the first to appear.

Eddy *et al.* (1958a,b), by adding 1% of inactivated calf serum to their mouse embryo tissue cultures, induced a greater host range in the virus. Hamsters injected at birth with this virus developed sarcomas mostly in the kidneys, but also in heart, lungs, liver, and other organs including the skin. Some of the tumours, especially those of liver and lungs were haemangiomas. These observations were amply confirmed by other groups of workers. Eddy *et al.* (1959a) also found rats to be susceptible to this virus; as with the hamsters the predominant tumours were sarcomas of the kidneys, but they too had multiple tumours including haemangiomas of the liver. Harris *et al.* (1961) infected ferrets, which developed fibro-sarcomas at the injection site; one tumour in a ferret was an osteo-sarcoma in the diaphragm with metastases to the liver and lungs. Eddy *et al.* (1959b) reported the induction of tumours in rabbits and (1960) in guinea pigs.

A number of workers have studied the maturation of polyoma virus within the cells it infects, giving an insight into the cell/virus relationships which lead to the development of cancer (for example – Negroni *et al.*, 1959; Vogt and Dulbecco, 1960, 1962; Dulbecco and Vogt, 1960; Eddy, 1960). In tissue cultures, the virus behaves for the most part as a normal non-carcinogenic virus; the virus multiplies in the usual way and the invaded cells are destroyed. In a few cells some stimulus is exerted, which leads to neoplastic transformation, more frequently in the alien hamster than in the mouse cells. The transformed cells do not release infective virus.

Rowe *et al.* (1961) showed that all strains of polyoma virus, even those from wild mice, are serologically identical. Neverthless different strains vary greatly in their potency, the tissues they attack and the form of tumour they induce (Dawe *et al.*, 1959; Hartley *et al.*, 1959; Sachs and Medina, 1960). All attempts to segregate the virus into different viral entities have failed, and the conclusion must be accepted that a single pluripotent virus is involved. The virus is, however, distinct from other members of the papovavirus group (Crawford and Crawford, 1963). Virus particles have been demonstrated both in tumour tissue and cultures by means of electron microscopy by a number of workers.

It remains in this chapter to give an account of the Shope Rabbit Papilloma Virus. Other members of the papovavirus group will be discussed in the appropriate place, together with the similar and probably related adenoviruses. They are all potentially important to the search for viruses as the causes of human cancers and their activities are of considerable interest.

V The rabbit papillomas

Mouse polyoma serves as a useful introduction to the world of the very small carcinogenic viruses of the papovavirus group, which are associated with warts and papillomas. Few groups of animals, if any, are free from them. Some of these viruses cause a mild type of disease accompanied by swellings of the skin. The tumours are generally benign and self-limiting, their progress being halted when they reach a certain size or it may be reversed so that the tumour disappears. However, some tumours in some hosts become secondarily malignant after an interval of time causing serious and often fatal neoplasia. The benign tumours in the natural host readily yield infective virus; this cannot be recovered by filtration from tumours that have become malignant though the presence of viral DNA can still be demonstrated. These facts, which have emerged during the course of extensive investigations, provide a fruitful field for the study of the mechanisms by which viral infections cause normal cells to become transformed into malignant ones. Polyoma virus itself, as we have seen, shows this dichotomy, since some infected cells are destroyed and shed infective virus, whereas others become transformed and do not.

Richard Shope (1933) of the Rockefeller Institute investigated naturally occurring skin tumours of wild cottontail rabbits in Kansas and Iowa. The wild rabbits are frequently affected by these tumours, which resemble large horny warts; sometimes the lesions appear as 'horns' on the side of the head or like rhinoceros horn on top of the nose. Shope demonstrated that the tumours could be transmitted by filtered material applied to the scarified skin of either domestic or wild cottontail rabbits, but no tumours resulted from introduction of virus elsewhere. The malignant potential of these viruses was demonstrated by Rous and Beard (1934a,b) by means of transplant techniques in domestic rabbits. The implanted papillomas acquired invasive properties, growing into the surrounding tissues and forming squamous cell carcinomas. They also found (Rous and Beard, 1935) that some of the subcutaneous papillomas induced by filtrates in domestic rabbits changed spontaneously into carcinomas. Syverton and Berry (1935) also observed a change from papilloma to carcinoma in a wild cottontail. Later Syverton *et al.* (1950) as a result of an extensive study of wild cottontails discovered that this change was not uncommon in them. Syverton (1952) reported further that more than a third of spontaneous tumours in wild cottontails had disappeared after six months, but none of the experimentally induced tumours did so. When naturally or experimentally infected, 25% of tumours in cottontails became transformed into carcinomas. Amongst domestic rabbits kept for 6 months, 75% of the tumours became malignant. From such tumours, no virus could be isolated. Electron microscope studies have confirmed that the virus belongs to the papovavirus group.

Papillomas of dogs, horses, cattle and man (including warts) have also been shown to belong to this same group of viruses. The bovine papilloma virus causes sarcoma-like tumours, when filtrates are transferred to horses and when injected into the brains of hamsters. It also 'transforms' both mouse and calf cells in culture. There is no evidence so far of malignancy associated with the papillomas of other speices. There is, however, no doubt that these viruses possess some malignant potential and are important in studies of the viral causes of cancer. The very fact that they are somewhat reluctant carcinogens gives scope for studying the conditions in which the cell/virus relationship leads to malignancy.

With the tardy recognition that the more important cancers of mice, like those of chickens, were the direct result of infections with filterable viruses, the search for similar causes in other animals and man became intensified. The course and results of these studies will be followed in ensuing chapters.

References

Aaronson, S. A. (1971). Isolation of a rat-tropic helper virus from M-M SV-O stocks. *Virology* **44**, 29–36

Aaronson, S. A. and Todaro, G. S. (1970). Transformation and virus growth by Murine Sarcoma Virus in human cells. *Nature* **225**, 29–36

Andervont, H. B. (1941). Effect of ingestion of strain C3H milk in the production of mammary tumours in strain C3H mice of different ages. *J. nat. Cancer Inst.* **2**, 13–16

Andervont, H. B. (1945). Susceptibility of young and adult mice to the mammary tumour agent. *J. nat. Cancer Inst.* **5**, 397–401

Andervont, H. B. and Bryan, W. R. (1944). Properties of the mouse mammary tumour agent. *J. nat. Cancer Inst.* **5**, 143–149

Bagg, H. J. (1925). The functional activity of the breast in relation to mammary carcinoma in mice. *Proc. Soc. exp. Biol. Med.* **22**, 419–421

Bentzvelen, P. (1974). Comparative biology of murine and avian tumour viruses. *In* (Kurstak, E. and Maramorosch, K., eds) 'Viruses, Evolution and Cancer'. New York and London: Academic Press

Bernard, C., Boiron, M. and Lasneret, J. (1967). Transformation et infection chronique de cellules embryonnaires de rat par le virus de Moloney. *CR Acad. Sci.* **264**, 2170–2173

Bernard, C., Lasneret, J., Boucher, M. and Boiron, M. (1969). Conversion cellulaire morphologique et replication virale après infection in vitro de cellules humaines per le virus du sarcome murin, souche Moloney. *CR Acad. Sci.* **268**, 624–627

Bernard, C., Chuat, J. C., Laprevotte, I. and Boiron, M. (1972). Further studies on mouse sarcoma virus (Moloney): replication in human cells. *Int. J. Cancer* **10**, 518–526

Bernhard, W. and Bauer, A. (1955). Mise en évidence de corpuscules d'aspect virusal dans des tumeurs mammaires de la souris. Étude au microscope électronique. *CR Acad. Sci.* **240**, 1380–1382

Bittner, J. J. (1936). Some possible effects of nursing on the mammary gland tumour incidence in mice. *Science,* **84**, 162

Bittner, J. J. (1939). Relation of nursing to the extra-chromosomal theory of breast cancer in mice. *Am. J. Cancer* **35**, 90–97

Bittner, J. J. (1940a). Breast cancer in mice as influenced by nursing. *J. nat. Cancr Inst.* **1**, 155–168

Bittner, J. J. (1940b). Further studies on native milk influence in breast cancer production in mice. *Proc. Soc. exp. Biol. Med.* **45**, 805–810

Bittner, J. J. (1942). Milk influence of breast tumours in mice. *Science* **95**, 462–463

Bittner, J. J. (1952). Transfer of the agent for mammary cancer by the male. *Cancer Res.* **12**, 387–398

Burrows, H. and Cook. J. W. (1936). Spindle-cell tumours and leukaemia in mice after infection with a water-soluble compound of 1:2:5:6-dibenzanthracene. *Am. J. Cancer* **27**, 267–278

Chesterman, F. C., Harvey, J. J., Dourmashkin, R. R. and Salaman, M. H. (1966). The pathology of tumours and other lesions induced in rodents by virus derived from a rat with Moloney leukaemia. *Cancer Res.* **26**, 1759–1768

Crawford, L. V. and Crawford, E. M. (1963). A comparative study of polyoma and papilloma viruses. *Virology* **21**, 258–263

Dawe, C. J., Law., L. W. and Dunn, T. B. (1959). Studies of parotid-tumor agent in cultures of leukaemic tissues of mice. *J. nat. Cancer Inst.* **23**, 717–797

Dmochowski, L. (1954). Discussion in: 'Proceedings, Symposium on 25 years of progress in mammalian genetics and cancer'. *J. nat. Cancer Inst.* **15**, 785–787

Dmochowski, L. and Passey, R. D. (1952). Attempts at tumor virus isolation. *Ann. NY Acad. Sci.* **54**, 1035–1066

Dobrovolskaia-Zavadskaia, N. (1933). Heredity of cancer susceptibility in mice. *J. Genetics* **27**, 191–198

Dulaney, A. D., Maxey, M., Schillig, M. G. and Goss, M. F. (1957). Neoplasms in C3H mice which received Ak leukaemic extracts when newborn. *Cancer Res.* **17**, 809–814

Dulbecco, R. and Vogt, M. (1960). Significance of continued virus production in tissue cultures rendered neoplastic by polyoma virus. *Proc. nat. Acad. Sci. USA* **46**, 1617–1623

Eddy, B. E. (1960). The polyoma virus. *In*: 'Advances in Virus Research', Section B, **46**, 1617–1623. New York: Academic Press

Eddy, B. E., Stewart, S. E. and Grubbs, G. E. (1958a). Influence of tissue culture passage, storage, temperature and drying on viability of SE polyoma virus. *Proc. Soc. exp. Biol. Med.* **99**, 289–292

Eddy, B. E., Stewart, S. E., Young, R. D. and Mider, G. B. (1958b). Neoplasms in hamsters induced by mouse tumor agent passed in tissue culture. *J. nat. Cancer Inst.* **20**, 747–761

Eddy, B. E., Stewart, S. E., Kirschstein, R. L. and Young, R. D. (1959a). Induction of subcutaneous nodules in rabbits with the SE polyoma virus. *Nature* **183**, 766–767

Eddy, B. E., Stewart, S. E., Stanton, M. F. and Marcotte, J. M. (1959b). Induction of tumors in rats by tissue culture preparations of SE polyoma virus. *J. nat. Cancer Inst.* **22**, 161–171

Eddy, B. E., Borman, G. S., Kirschtein, R. L. and Touchette, R. H. (1960). Neoplasms in guinea pigs infected with SE polyoma virus. *J. infect. Dis.* **107**, 361–368

Finkel, M. P., Birkis, B. O. and Jinkins, P. B. (1966). Virus induction of osteosarcomas in mice. *Science* **151**, 698–700

Friend, C. (1956). The isolation of a virus causing malignant disease of the haematopoietic system in adult Swiss mice. (abstract) *Proc. Am. Assoc. Cancer Res.* **2**, 106

Friend, C. (1957). Cell-free transmission in adult Swiss mice of a disease having the character of a leukaemia. *J. exp. Med.* **109**, 217–228

Furth, J. (1951). Recent studies on the nature and etiology of leukaemia. *Blood* **6**, 964–975

Furth, J. (1952). Recent experimental studies and current concepts on the etiology and nature of leukaemia. *Proc. Inst. Med. Chicago* **19**, 95–104

Furth, J. and Furth, O. B. (1938). Monocytic leukaemia and other neoplastic diseases occurring in mice following intra-splenic injection of 1:2 benzpyrene. *Am. J. Cancer.* **34**, 169–183

Furth, J. and Strumia, M. (1931). Studies on transmissible lymphoid leukaemia of mice. *J. exp. Med.* **53**, 715–731

Furth, J., Seibold, H. R. and Rathbone, R. R. (1933). Experimental studies on lymphomatosis in mice. *Am. J. Cancer* **19**, 521–604

Furth, J., Buffett, R. F., Banasiewicz-Rodriguez, M. and Upton, A. C. (1956). Character of agent including leukaemia in newborn mice. *Proc. Soc. exp. Biol. Med.* **93**, 165–172

Gardner, W. U. (1937). 'Influence of estrogenic hormones on abnormal growths'. Occas. Public. Am. Assoc. Adv. Sci., pp. 67–75. USA: Science Press

Gardner, W. U., Kirschbaum, A. and Strong, L. C. (1940). Lymphoid tumors in mice receiving estrogens. *Arch. Path.* **29**, 1–7

Gardner, W. U., Dougherty, T. F. and Williams, W. L. (1944). Lymphoid tumors in mice receiving steroid hormones. *Cancer Res.* **4**, 73–87

Graffi, A. (1957). Chloroleukaemia of mice. *Ann. NY Acad. Sci.* **68**, 540–558

Graffi, A., Moore, D. N., Stanley, W. M., Randall, H. T. and Haagensen, C. D. (1949). Isolation of mouse mammary carcinoma virus. *Cancer* **2**, 755–762

Gross, L. (1951a). 'Spontaneous' leukaemia developing in C3H mice following inoculation, in infancy, with Ak leukaemic extracts, or Ak embryos. *Proc. Soc. exp. Biol. Med.* **76**, 27–32

Gross, L. (1951b). Pathogenic properties and 'vertical' transmission of the mouse leukaemia agent. *Proc. Soc. exp. Biol. Med.* **78**, 342–348

Gross, L. (1953). A filterable agent recovered from Ak leukaemic extracts causing salivary gland carcinomas in C3H mice. *Proc. Soc. exp. Biol. Med.* **78**, 342–348

Gross, L. (1955). Induction of parotid carcinomas and/or subcutaneous sarcomas in C3H mice with normal C3H organ extracts. *Proc. Soc. exp. Biol. Med.* **88**, 362–368

Gross, L. (1963). Serial cell-free passage in rats of the mouse leukaemia virus. Effect of thymectomy. *Proc. Soc. exp. Biol. Med* **112**, 939–945

Gross, L. (1970). 'Oncogenic Viruses', 2nd edn, 991pp. Oxford: Pergamon Press

Gross, L., Gessler, A. E. and McCarty, A. S. (1950). Electron microscopic examination of human milk particularly from women having family record of breast cancer. *Proc. Soc. exp. Biol. Med.* **75**, 270–276

Gross, L., McCarty, A. S. and Gessler, A. E. (1952). The significance of particles in human milk. *Ann. NY Acad. Sci.* **54**, 1018–1034

Gross, L., Roswit, B., Malsky, S. J., Dreyfuss, Y. and Amato, C. G. (1965). Resistance of mouse leukaemia virus to *in vitro* gamma ray irradiation. (abstract) *Proc. Am. Assoc. Cancer Res.* **6**, 24

Guérin, M. (1955). Corps d'inclusion dans les adénocarcinomes mammaires de la souris. *Bull. du Cancer* **42**, 14–28

Guérin, M. (1956). Influence des rayons sur l'apparition d'adénocarcinomes mammaires chez des souris inoculées avec de tumeurs contenants des corps d'inclusions. *Bull, du Cancer* **43**, 23–36

Harris, R. J. C., Chesterman, F. C. and Negroni, G. (1961). Induction of tumours in newborn ferrets with Mill Hill polyoma virus. *Lancet* **1**, 788–791

Hartley, J. W. and Rowe, W. P. (1966). Production of altered cell foci in tissue culture by defective Moloney sarcoma virus. *Proc. nat. Acad. Sci. USA* **55**, 780–786

Hartley, J. W., Rowe, W. P., Chanock, R. M. and Andrews, B. E. (1959). Studies of mouse polyoma virus infection. IV Evidence for mucoprotein erythrocyte receptors in polyoma virus haemagglutination. *J. exp. Med.* **110**, 81–91

Hartley, J. W., Rowe, W. P., Capps, W. I. and Huebner, R. J. (1965). Complement fixation and tissue culture assays for mouse leukaemia viruses. *Proc. nat. Acad. Sci. USA* **53**, 931–938

Harvey, J. J. (1964). An unidentified virus which causes the rapid production of tumours in mice. *Nature* **204** 1104–1105

Harvey, J. J. (1969). Susceptibility of *Praomys* (*Mastomys*) *natalensis* to the murine sarcoma virus – Harvey (SV-H). *Int. J. Cancer* **3**, 634–643

Harvey, J. J. and East, J. (1969). Biological activity and separation of a leukaemogenic virus from murine sarcoma virus-Harvey (MSV-H). *Int. J. Cancer* **4**, 655–665

Harvey, J. J., Salaman, M. H., Chesterman, F. C., Gillespie, A. V., Harris, R. J. C., Evans, R. and Mahy, B. W. J. (1964). 'Studies on a murine sarcoma virus (MSV)' Brit. Empire Cancer Campaign for Research, 42nd Annual Report (Part II), pp. 185–189

Hays, E. F. and Beck, W. S. (1958). The development of leukaemia and other neoplasms in mice receiving cell-free extracts from a high-leukaemia (AKR) strain. *Cancer Res.* **18**, 676–681

Huebner, R. J., Hartley, J. W., Rowe, W. P., Lane, W. T. and Capps, W. I. (1966). Rescue of the defective genome of Moloney sarcoma virus from a non-infectious hamster tumor and the production of pseudotype sarcoma viruses with various murine leukaemia viruses. *Proc. nat. Acad. Sci. USA* **56**, 1164–1169

Ida, N. (1957). Discussion following paper: 'Filterable agent causing leukaemia following inoculation into newborn mice', by L. Gross. *Texas Rep. Biol. Med.* **15**, 616–618

Kaplan, H. S. (1967). On the natural history of the murine leukaemias. *Cancer Res.* **27**, 1325–1340

Kawamoto, S., Ida, N., Kirschbaum, A. and Taylor, G. (1958). Urethan and leukaemogenesis in mice. *Cancer Res.* **18**, 725–729

Kinosita, R., Erickson, J. O., Armen, D. M., Dolch, M. E. and Ward, J. P. (1953). Electron microscope study of mouse mammary carcinoma tissue. *Exp. Cell Res.* **4**, 353–361

Kirschbaum, A. (1957a). Genetic and nongenetic factors influencing the induction of mouse leukaemia. *In* 'The Leukaemias, Etiology, Pathophysiology, and Treatment'. C6, pp. 121–125. New York: Academic Press

Kirschbaum, A (1957b). Discussion following paper: 'Filterable agent causing leukaemia following inoculation into newborn mice', by L. Gross. *Texas Rep. Biol. Med.* **15**, 618–620

Kirschbaum, A., Shapiro, J. R. and Mixer, H. W. (1953). Synergistic action of leukemogenic agents. *Cancer Res.* **13**, 262–268

Kirsten, W. H. and Mayer, L. A. (1967). Morphological responses to a murine erythroblastosis virus. *J. nat. Cancer Inst.* **39**, 311–335

Kirsten, W. H., Somers, K. D. and Mayer, L. A. (1968). Multiplicity of cell response to a murine erythroblastosis virus. *In* 'Proc. 3rd Internat. Symposium on Comparative Leukaemia Research, Paris (July, 1967)'. *Bibl. Haemat.* No. 30, pp. 64–65. Basel and New York: Karger

Korteweg, R. (1934). Profondervindelijke onderzoekingen aangaande erfelijkheid van kanker. *Nederland, Tydschr. v. geneesk.* **78**, 240–245

Korteweg, R. (1936). On the manner in which the disposition to carcinoma is inherited in mice. *Genetics* **18**, 350–371

Lacassagne, A. (1937). Sarcome lymphoides chez des souris loguement traités per des hormones oestrogènes. *CR Soc. Biol.* **126**, 193–195

Lasfargues, E. Y. (1964). Etiologie virale des tumeurs mammaires de la souris. *Laval médical.* **35**, 901–908

Lathrop, A. E. C. and Loeb, L. (1918). Further investigation on the origin of tumours in mice. V The tumour rate in hybrid strains. *J. exp. Med.* **28**, 475–500

Law, L. W. (1954). Recent advances in experimental leukaemia research. *Cancer Res.* **14**, 695–709

Law, L. W., Dunn, T. B. and Boyle, P. J. (1955). Neoplasma in the C3H strain and in F1 hybrid mice of two crosses following the introduction of extracts and filtrates of leukaemic tissues. *J. nat. Cancer Inst.* **16**, 495–539

Lieberman, M. and Kaplan, H. S. (1959). Leukemogenic activity of filtrates from radiation-induced lymphoid tumours of mice. *Science* **130**, 387–388

Little, C. C. (1947). The genetics of cancer in mice. *Biol. Rev.* **22**, 315–343

Little, C. C. (1958). Biological aspects of cancer research. *J. nat. Cancer Inst.* **20**, 441–464

Marsh, M. C. (1929). Spontaneous mammary cancer in mice. *J. Cancer Res.* **13**, 313–339

McDowell, E. C. and Richter, M. N. (1935). Mouse leukaemia. IX The role of heredity in spontaneous cases. *Arch. Path.* **20**, 709–724

Moloney, J. B. (1959). Preliminary studies on a mouse lymphoid leukaemia virus extracted from Sarcoma 37. (abstract) *Proc. Am. Assoc. Cancer Res.* **3**, 44

Moloney, J. B. (1960). Biological studies on a lymphoid leukaemia virus extracted from sarcoma 537. I origin and introductory investigation. *J. nat. Cancer Inst.* **24**, 933–951

Moloney, J. B. (1966a). 'A virus-induced rhabdomyosarcoma in mice.' Conference on murine leukaemia. Nat. Cancer Inst. Monograph no. 22, pp. 139–142. US Public Health Service, Bethesda Md.

Moloney, J. B. (1966b). The application of studies in murine leukaemia to the problems of human neoplasia. *In* (Fiennes, R. N., ed.) 'Some Recent Developments in Comparative Medicine'. Symposium of the Zoological Society of London, no. 17, pp. 251–258. London and New York: Academic Press

Moore, D. N. (1963). Mouse mammary tumour agent and mouse mammary tumours. *Nature* **198**, 429–433

Murray, W. S. (1928). Ovarian secretion and tumour incidence. *J. Cancer Res.* **12**, 18–25

Negroni, G., Dourmashkin, R. and Chesterman, F. C. (1959). A 'polyoma' virus derived from a mouse leukaemia. *Brit. Med. J.* **2**, 1359–1360

Passey, R. D., Dmochowski, L., Astbury, W. T., Reed, R. and Johnson, D. (1950). Electron microscope studie of normal and malignant tissues of high- and low-breast cancer strains of mice. *Nature* **165**, 107

Porter, K. R. and Thompson, H. P. (1948). A particulate body associated with epithelial cells cultured from mammary carcinoma of mice of a milk factor strain. *J. exp. Med.* **88**, 15–24

Pybus, F. C and Miller, E. W. (1934). Hereditary mammary carcinomas of mice (a description of 100 consecutive tumours). *Newcastle med. J.* **14**, 151–169

Pybus, F. C. and Miller, E. W. (1938). Multiple sarcomas in a sarcoma strain of mice. *Am. J. Cancer* **34**, 252–254

Pybus, F. C. and Miller, E. W. (1940). The gross pathology of spontaneous bone tumours in mice. *Am. J. Cancer* **40**, 47–61

Rauscher, F. J. (1962). A virus induced disease of mice characterized by erythrocytopoiesis and lymphoid leukaemia. *J. nat Cancer Inst.* **29**, 515–543

Richter, M. N. and McDowell, E. C. (1929). The experimental transmission of leukaemia in mice. *Proc. Soc. exp. Biol. Med.* **26**, 362–364

Richter, M. N. and McDowell, E. C. (1930). Studies on leukaemia in mice. I The experimental transmission of leukaemia. *J. exp. Med.* **51**, 659–673

Rous, P. and Beard, J. W. (1934a). Carcinomatous changes in virus-induced papillomas of the skin of the rabbit. *Proc. Soc. exp. Biol. Med.* **32**, 578–580

Rous, P. and Beard, J. W. (1934b). A virus induced growth with the characters of a tumour (the Shope rabbit papilloma). I The growth on implantation within favorable hosts. *J. exp. Med.* **60**, 701–722

Rous, P. and Beard, J. W. (1935). The progression to carcinoma of virus-induced rabbit papillomas (Shope). *J. exp. Med.* **62**, 523–548

Rowe, W. P., Huebner, R. J. and Hartley, J. W. (1961). An approach to the study of tumor viruses. Ecology of a mouse tumor virus. *In* 'Perspectives in Virology', Vol. 2. Minneapolis: Burgess Publishing Co.

Sachs, L. and Medina, D. (1960). Polyoma virus mutant with a reduction in tumour formation. *Nature* **184**, 1702–1704

Shope, R. E. (1933). Infectious papillomatosis of rabbits. *J. exp. Med.* **58**, 607–624

Simons, P. J., Bassin, R. H. and Harvey, J. J. (1967). Transformation of hamster embryo cells *in vitro* by murine sarcoma virus (Harvey). *Proc. Soc. exp. Biol. Med.* **125**, 1242–1246

Smith, G. H. (1965). The role of the milk agent in the disappearance of mammary tumours in inbred C3H/SiWi mice. (abstract) *Proc. am. Assoc. Cancer Res.* **6**, 60

Stewart, S. E. (1953). Leukaemia in mice produced by a filterable agent present in AKR leukaemic tissues with notes on a sarcoma produced by the same agent. (abstract) *Anat. Rec.* **117**, 532

Stewart, S. E. (1955). Neoplasms in mice inoculated with cell-free extracts or filtrates of mouse leukaemic tissues. I Neoplasms of the parotid and adrenal glands. *J. nat. Cancer Inst.* **15**, 1391–1415

Stewart, S. E., Eddy, B. E., Gochenour, A. M., Borghese, N. G. and Grubbs, G. E. (1957). The induction of neoplasms with a substance released from mouse tumours by tissue culture. *Virology* **3**, 380–400

Stewart, S. E., Eddy, B. E. and Borghese, N. (1958). Neoplasms in mice inoculated with a tumor agent carried in tissue culture. *J. nat. Cancer Inst.* **20**, 1223–1243
Strong, L. C. (1935). The establishment of the C3H inbred strain of mice for the study of spontaneous carcinoma of the mammary gland. *Genetics* **20**, 580–591
Strong, L. C. (1936). The establishment of the 'A' strain of inbred mice. *J. Heredity* **27**, 21–24
Strong, L. C. (1942). The origin of some inbred mice. *Cancer Res.* **2**, 531–539
Syverton, J. T. (1952). The pathogenesis of the rabbit papilloma-carcinoma sequence. *Ann. NY Acad. Sci.* **54**, 1126–1140
Syverton, J. T. and Berry, G. P. (1935). Carcinoma in the cottontail rabbit following spontaneous virus papilloma (Shope). *Proc. Soc. exp. Biol. Med.* **72**, 46–50
Syverton, J. T., Dascomb, H. E., Wells, E. B., Koomen, J. jr. and Berry, G. P. (1950). The virus induced papilloma-carcinoma sequence. II Carcinomas in the natural host, the cottontail rabbit. *Cancer Res.* **10**, 440–444
Thomas, M., Boiron, M., Stoytchkov, Y. and Lasneret, J. (1968). *In vitro* replication of Mouse Sarcoma Virus (Moloney strain) in bovine embryo skin cells. *Virology* **36**, 514–518
Ting, R. C. (1966). *In vitro* transformation of rat embryo cells by a murine sarcoma virus. *Virology* **28**, 783–785
Vogt, M. and Dulbecco, R. (1960). Virus-cell interaction with a tumor-producing virus. *Proc. nat. Acad. Sci. USA* **46**, 365–370
Vogt, M. and Dulbecco, R. (1962). Studies on cells rendered neoplastic by polyoma virus. The problem of the presence of virus-related materials. *Virology* **16**, 41–51
Woolley, G. W. and Small, M. C. (1956). Experiments on cell-free transmission of mouse leukaemia. *Cancer* **9**, 1102–1106

6

Tumour Systems in Cats and Cattle

I General

After fifty years of endeavour, two major tumour systems were known to be caused by viral infections, the leukosis and sarcoma complexes of chickens and mice. A further system, the hormone-dependent mouse mammary carcinoma had also been proved to be of viral origin. The group of viruses involved in these cancers was assigned to a special group, which they named *oncornavirus*. There are a number of sub-groups of oncornavirus, known as A, B, C, D and E. These viruses are known as 'RNA Viruses', because they contain only RNA, no DNA. There is thus a distinction between the RNA virus-induced cancers and the DNA virus-induced cancers. The DNA cancer viruses belong to three groups, the *Papovavirus,* the *Adenovirus* and the *Herpesvirus* groups. We have not yet encountered cancers caused by adenoviruses, which will be considered with the human and non-human primate transmissible cancers.

The DNA viruses replicate in the cell nucleus, the RNA viruses in the cytoplasm attached to the cell membranes. Before they can do this, however, they need to undergo a preliminary stage of development in the nucleus in the form of a 'provirus'. Since RNA cannot be accepted into the nucleus, they become 'transcripted' into DNA copies of themselves by an enzyme, which they produce, known as 'reverse transcriptase'. In the nucleus, the first stage of development proceeds and the proviral forms are returned as RNA to the cytoplasm, where mature 'virions' are manufactured. The infectious process then proceeds in the normal way by destruction of the parent cell and release of virus particles ('virions') to infect fresh cells or be destroyed by the body defences. This does not, however, always happen. These viruses have the extraordinary ability, when in the nucleus, to displace some host genes and replace them with their own in the host genetic material. When this happens, mature oncornavirus cannot be detected because it is not there. Under certain conditions, the viral genes may be stimulated to produce mature virus again; under other conditions, the viral nucleotides may exert a malignant influence on the cell, which may lead to the cell becoming neoplastic.

In a healthy natural host, the resident oncornavirus is likely to do neither of these things. Its genes simply remain in the host's genetic apparatus and behave as if they were host genes. When the host's chromosomes divide preparatory to forming two daughter cells, the viral genes also divide and pass into the daughters. If they are present in the reproductive cells, whether sperm or egg, they are transmitted to the embryo. In this situation, they do not normally cause cancer, unless the host cell is damaged by carcinogenic chemicals, ionising radiation, or in some other way, or unless the provirus is transferred by some means to a host other than the natural one. The harmless presence of these oncornavirus genes can be detected in all cells of all vertebrates from fish to man. Moloney (1976) has expressed the situation as follows:

> It appears that RNA viruses are a normal part of all cells of all vertebrates and it is likely that they function in embryonic development and only under certain conditions in oncogenesis.

These viruses, then, are always with us and have been for a very long time, being transmitted from parents to progeny during thousands of millions of years of evolution, evolving themselves *pari passu* with their hosts. It has even been suggested that they play a beneficial role in evolution itself.

However, not all oncornaviruses are transmitted in this way and this adds to the complexity of the viral cancer problem. Moloney (1975) separates the oncornaviruses on the basis of their natural hosts and behaviour into three categories: 1. endogenous xenotropic type C viruses, which are mostly

Table 1 Endogenous xenotropic type-C viruses. (After Moloney, 1975)

Virus strain		Investigators
Mouse	NZB	Levy & Pincus (1970)
	NIH (Swiss)	Todaro *et al.* (1973)
	BALB/c 2	Stephenson *et al.* (1974)
	C57 L	Arnstein *et al.* (1974)
	C57 BL/6	Benveniste *et al.* (1974)
Hamster	HALV-1	Freeman *et al.* (1974)
Rat	RALV-1	Klement *et al.* (1973)
Cat	RD 114	McAllister *et al.* (1972); Fischinger *et al.* (1972); Livingston & Todaro (1973); Sarma *et al.* (1973)
Chicken	Strain E	Weiss *et al.* (1973)
Swine*	PK 15	Moennig *et al.* (1974); Todaro *et al.* (1974)
Baboon*	Bab LV	Kalter *et al.* (1973); Benveniste *et al.* (1974); Todaro *et al.* (1974)
Human ?*	Baboon-virus related	Sherr & Todaro (1974)

*Animals in which xenotropic virus is only type demonstrated or suspected.

Table 2 Endogenous ecotropic viruses. (After Moloney, 1975)

Virus strain		Investigators
Mouse	AKR Strains	Gross (1961)
	C57 BL/6-RADLV	Lieberman & Kaplan (1959)
	BALB/c-N-tropic	Aaronson *et al.* (1969); Hartley *et al.* (1970)
	BABB/c-B-tropic	Hartley *et al.* (1970)
	MTV milk factor	Bittner (1936)
Wild mouse	LC & 1504 E	Gardner *et al.* (1971)
Hamster	HALV-2 MSV HALV	Kelloff *et al.* (1970)
Rat	RALV-2 MSV (O)	Ting (1968); Bergs *et al.* (1972)
Cat	FeLV 1, 2, 3 FSV	Theilen *et al.* (1969); Rickard *et al.* (1969); Gardner *et al.* (1970); Sarma *et al.* (1970)

genetically transmitted and do not harm their natural hosts, though they may cause pathological changes in the cells of other species, 2. endogenous ecotropic viruses, which may be transmitted genetically or epigenetically (as in the mother's milk) and which are pathogenic both for the cells of the natural host and possibly for those of some other species as well; and 3. exogenous RNA tumour viruses, which are very pathogenic for the natural host though not normally resident; they are often pathogenic also for heterotypic hosts. Reproduction of Moloney's tables (Tables 1–3) will illustrate the more important animals, from which each category of viruses has been isolated, and will also demonstrate some of the difficulties, which face investigators.

Oncornaviruses and oncornavirus-like particles have been observed in a

Table 3 Exogenous RNA tumour viruses. (After Moloney, 1975)

	Investigators
1. *Vertically transmitted (i.e. genetic)*	
ALV (avian leukosis viruses A, B, C, D)	Rubin *et al.* (1961)
MuLV (murine leukosis viruses GLV FLV MLV RLV)	Gross (1951, 1958); Friend (1957); Moloney (1960); Rauscher (1962)
2. *Horizontally transmitted through excretions or milk*	
Chicken ALV A & B	Burmester & Waters (1955); Beard (1957)
Mouse mammary tumour virus (MMTV)	Bittner (1936); Andervont & Dunn (1948
Cat feline leukaemia (FeLV)	Hardy *et al.* (1973)
Gibbon ape & woolly monkey viruses	Kawakami *et al.* (1972); Parks *et al.* (1973); Gilden *et al.* (1974)

great many species or in cultured cells derived from them, but their association with cancers has yet to be proved in a number of them. Such species include: man (Dmochowski *et al.*, 1965; Priori *et al.*, 1971; McAllister *et al.*, 1972), monkeys (Theilen *et al.*, 1971; Wolfe *et al.*, 1971; Kalter *et al.*, 1973); cattle (Miller *et al.*, 1969; Stock and Ferrer, 1972); swine (Howard *et al.*, 1968); guinea pigs (Nadel *et al.*, 1967; Hsiung, 1972); rats (Weinstein and Moloney, 1965; Klement *et al.*, 1971); hamsters (Stenback *et al.*, 1968; Kelloff *et al.*, 1970); and snakes (Ziegel and Clark, 1969). Evidence of 'viral markers' has been found in every case, where they have been sought. Some RNA viruses exist primarily as infectious particles, but most are in the repressed form integrated into the cellular chromosomes. Their presence in no way indicates the existence of any form of disease, though they may cause such if derepressed by carcinogenic chemicals or radiation. They may never cause disease in their natural host, but show their oncogenic potential by causing disease in other animals. They are potentially dangerous, too, because related RNA viruses readily exchange genes to form hybrid recombinants, which may be extremely virulent. Such has been frequently observed in the laboratory, and could account for the occurrence of serious cancers in man and his domestic animals.

The extraordinary life history of these viruses has been discovered as a result of the original isolation of viruses from chickens and mice. Until this knowledge had been acquired, the search for infectious causes of cancer would have been virtually hopeless. Meanwhile, much of the further advance in finding the viruses responsible for cancers in other animals has been dependent on this knowledge, and it has been necessary to outline it at this point. Although oncornavirus cancers have now been discovered and studied in a number of animals, the account here given will be confined to those of cats and cattle, each of which give fresh insights into these problems.

II Oncornavirus infection of cats

Cats suffer from a variety of leukaemic diseases, mostly lymphosarcoma but occasionally lymphatic and granulocytic (bone marrow) leukaemias; they also suffer from true sarcomas. According to Cotchin (1952, 1957) and Jarrett (1966), the leukoses account for 9%–15% of all cat tumours. The sarcomas, usually fibro-sarcomas, occur less frequently and in older animals; the leukaemias usually affect quite young animals.

The transmission of feline leukaemia by cell-free material, obtained by centrifugation, was first achieved by Jarrett *et al.* (1964b) at the Glasgow Veterinary School. The material was taken from tumour tissue of an 8½-year-old female cat with fibro-sarcomas affecting thymus, spleen and lymphatic glands. The prepared material was injected into four kittens of a litter less than twelve hours old. All the kittens developed disseminated lympho-sarcomas after 9–18 months. In the second passage, Jarrett (1966) reported that lympho-sarcomas appeared in only 8 weeks. Rickard *et al.* (1967) transmitted feline leukaemia by means of cell suspensions, that is non-cell-free material. Transmission of feline sarcomas was achieved by means of filtrates by Snyder and Theilen (1969). The material was obtained from multiple subcutaneous fibro-sarcomas in a 2-year-old female Siamese cat and injected into newborn kittens. The tumours could be serially transmitted appearing in the short time of 2–3 weeks. This virus was tested also in dogs, but failed to infect them or cause tumours, which regressed; the same happened in rabbits and rhesus monkeys.

The feline leukaemia viruses have been extensively studied since their discovery, by Kawakami *et al.* (1967), in tissue cultures and by electron microscopy. They are typical type C oncornaviruses similar to those, which cause leukoses and sarcomas in chickens and mice, though there is no evidence so far that the sarcoma viruses are incomplete. They have been much used in studies of the infectious leukaemias, and readily hybridise with other viruses of the group. They possess one odd peculiarity, in that they are always transmitted horizontally; they do not ever appear to pass vertically by genetic transmission. Cats do, however, possess an endogenous oncornavirus, which is vertically transmitted, but this has not so far been associated with any of the leukaemias or other cancers. There are a number of serologically distinguishable sub-types of the feline virus, which may be associated with different forms of the disease. In kittens, there is a form of the disease in which the tumour arises from the thymus and comes to occupy much of the chest cavity. There is also an alimentary form of the disease, in which there are tumours of the intestines, mesenteric lymph nodes and spleen; tumours of the liver and kidneys are not uncommon and some cats develop lymphoid cells in the anterior chamber of the eye. There are occasionally lymphatic leukaemic changes in the blood, but the blood cells are usually normal. Jarrett *et al.* (1964a) and Jarrett (1966) failed to find viral particles in tissues from spontaneous cases, but these were present after the second tissue culture passage and in cultured cells. However, Laird *et al.* (1968a, b), working in Jarrett's laboratory, described their presence in a spontaneous case examined by them. They were also described by Rickard *et al.* (1968).

III Oncornavirus infection of cattle

The leukaemias of cattle resemble those of cats, though the virus is serologically distinct. Their study is of importance for four main reasons: 1. cattle provide a model for the study of these diseases in a large and potentially long-lived animal, in which they develop slowly after a long incubation period, the situation in this respect might be more relevant to that in human beings, 2. the tumours quite frequently regress and thus present opportunities for the study of cell/virus relationships and immune mechanisms, 3. the proof that cattle leukaemias were caused by viruses proved elusive, partly because of the long incubation period and the consequent difficulty of associating cause and effect; here again, there is a potential parallel with the situation as it may be in man, and 4. the real possibility that an effective vaccine can be produced, which will control the incidence of the disease (*vide* WHO, 1973; Miller *et al.*, 1972).

As with cats, there are three forms of the disease: 1. lympho-sarcoma of adult animals, 2. a cutaneous form of juveniles, and 3. a thymic form, which affects young adults. The juvenile form usually occurs at about 6 months, but may be present at birth; it is in this form that regression occurs. In cases, in which the thymus is involved, the animals are usually 1–2 years old and lymphatic glands may also be involved. The adult form of the disease is the commonest, occurring in cattle of 5–6 years; the tumours are found in scattered lymph nodes and organs. Both leukaemia and lympho-sarcoma have been on the increase in cattle in a number of countries in recent years; these include the United States, East Germany, Sweden and Denmark. The incidence figures overall are not unduly high, but there exist leukaemic herds and families of cattle, in which 2%–5% of the adults become affected with lympho-sarcoma. This has been reported by many workers over the years, as reviewed by Gross (1970). Particularly interesting case histories are given by Böttger (1954, 1955) collected from a number of farmers in Germany; in a number of these cases, there is a history of wide dissemination of the disease to the cows by infected herd bulls, and thence from the cows to the calves.

The herd and familial incidence of leukaemias and lympho-sarcomas was regarded as *a priori* evidence that they were of infectious origin. Attempts to eradicate the disease in Denmark and West Germany on this basis, using counts of white blood cells as evidence of pre-cancerous changes, were of limited success, but there were small reductions in disease incidence. Early attempts to transmit the disease by cell grafts, as by Marshak *et al.* (1967) gave disappointing results, because: 1. the length of time between the transplant and the appearance of tumours was too long for firm conclusions to be drawn, and 2. tumours arising in younger animals regressed. Attempts to transmit the disease by filtrates, as by Hoflund *et al.* (1963) and Olson (1961) were unsuccessful. Similarly attempts, as by Dutcher *et al.* (1964) to demonstrate viral particles by electron microscopy were inconclusive. A situation had arisen, in which there was epidemiological evidence that these diseases must be infectious, but it was not possible to prove it. However, Miller *et al.*

(1969) did eventually find C particles of oncornavirus in tissue cultures of blood lymphocytes from 9 of 12 cattle with lympho-sarcoma, and the viral cause of the disease was thereafter demonstrated by several workers (Ferrer *et al.*, 1971; Miller *et al.*, 1972; Olson *et al.*, 1973). It was also shown by Olson *et al.* (1972) that lympho-sarcoma could be readily transmitted to sheep and that the incubation period in them was relatively short. It has been shown also that transmission of the bovine leukoses can be either vertical or horizontal, unlike the feline. Serological tests have now been developed, which are very reliable in identifying animals carrying sub-clinical infections and those which have been infected and subsequently recovered; negative results are a reliable indication that the animal has never been exposed to infection. Ferrer (1972) was able to show by the use of serological tests that the bovine virus is antigenically distinct from the feline.

It is not proposed to discuss further the incidence of oncornavirus cancers in other animals. Suffice it to say that leukotic diseases are now known to be caused by infectious agents in all animals that have been studied, with the single exception of man. The situation with regard to the human disease resembles that which formerly existed with the bovine; viral particles have been demonstrated in tumour tissue and antibodies against them are present in affected patients. The position in non-human primates and man will be studied in the next two chapters. Meanwhile, we now study the bovine papillomas.

IV The bovine papillomas

The papillomas, which frequently affect the skins of cattle, are benign growths and have the character of warts. That they are transmissible by filtrates was demonstrated a long time ago by Magalhães (1920) in Brazil. The virus responsible has been studied by electron microscopy by Lévy *et al.* (1963) and proved to be similar to the Shope Papilloma Virus of cottontail rabbits and to the human wart virus. Often the warts regress spontaneously and disappear. Up to this point, both disease and virus are of little significance. However, there is another form of the disease, in which tumours develop in the bladder or vagina; some of them develop secondarily into carcinomas or haemangiomas. Sometimes, papillomas of the bladder develop in calves, which have received injections of virus obtained from lesions of the skin; conversely, virus obtained from bladder papillomas may cause either skin papillomas or bladder papillomas. It is, therefore, evident that the same virus is responsible for causing both types of lesion.

Campo *et al.* (1980) published in paper in *Nature* entitled 'A new papillomavirus associated with alimentary cancer in cattle'. As the title indicates, he and his colleagues at the Glasgow Veterinary School, including Jarrett and Laird, whose names are associated with the demonstration of the transmissible nature of feline leukaemia, isolated a virus of the papova group from bowel cancers in cattle. They attributed a causal role to the virus, but suggested that there might also be a link with poisoning by bracken (*Pteridium*). This discovery is potentially of importance, and it is to be hoped that bowel cancers of other animals, including man, will be examined for the presence of similar viruses. Bowel cancers are an especially intractable problem, because of their tendency to metastasise to other sites especially the liver, when treatment becomes virtually hopeless. This discovery could, if the etiological relationship of the virus with the development of the cancer were established, lead to important results.

The bovine papillomas are of importance for two reasons. In the first place, man too suffers from warts of the skin, which are benign and may regress spontaneously or disappear with treatment, even by 'wart charmers'; it is fully established from transmission experiments on human volunteers that the warts are caused by a filterable virus, similar to that which causes the bovine papillomas. Man, too, suffers from papillomas of the urinary bladder, which can usually be surgically removed with success, but which may also sometimes become malignant. The question arises as to whether man's bladder tumours, as with those of cattle, are also caused by his natural wart virus? The bovine papillomas present a useful model for studying the problem. Secondly, calves which recover from skin warts are immune to reinfection with the virus. Furthermore, a safe killed virus vaccine can be prepared. Here then is a second cancer, the first being Marek's Disease of chickens, the development of which can be prevented by means of a vaccine. The bovine virus is infective also for horses and hamsters, both of which may develop neoplastic conditions of the skin as a result of infection; the hamster tumours sometimes also develop metastases.

Papillomas occur naturally in many groups of animals, including horses, and dogs in which they frequently occur in the mouth. The subject need not be pursued further here. In the next chapter, we shall study the neoplastic diseases of monkeys and apes. The story is one of great interest and fascination, and is of importance because these animals are most closely related to man and what happens in them is most likely to be similar to what happens in man.

References

Aaronson, S. A., Hartley, J. W. and Todaro, G. J. (1969). Mouse leukaemia virus: spontaneous release by mouse embryo cells after long term *in vitro* cultivation. *Proc. nat. Acad. Sci. USA* **64**, 87–94

Andervont, H. B. and Dunn, T. B. (1948). Mammary tumors in mice presumably free of the mammary tumour agent. *J. nat. Cancer Inst.* **8**; 227–233

Arnstein, P., Levy, J. A., Oshiro, L. S., Price, P. J., Suk, W. A. and Lennette, E. H. (1974). Recovery of murine xenotropic type-C virus from C57/Leaden mice. *J. nat. Cancer Inst.* **53**, 1787–1792

Beard, J. W. (1957). Etiology of avian leukosis. *Ann. NY Acad. Sci.* **68**, 473–486

Benveniste, R. E., Heinemann, R., Wilson, G. L., Callahan, R. and Todaro, G. J. (1974). Detection of baboon type C viral particles in various primate tissues by molecular hybridization. *J. Virol.* **14**, 56–67

Bergs, V., Scotti, T. M. and Bergs, M. (1972). Rat leukaemia derived 9H virus (9HV): II Response of rats to low doses of virus. *Proc. Soc. exp. Biol. Med.* **139**, 535–539

Bittner, J. J. (1936). Some possible effects of nursing on the mammary gland tumor incidence in mice. *Science* **84**, 162

Böttger, T. (1954). Beobachtungen über das Auftreten und die Erblichkeit der tumorösen Form der Rinderleukose. *Zeitschr. f. Tierzucht u. Züchtungsbid.* **63**, 223–238

Böttger, T. (1955). Weitere Ermittlungen über das Auftreten und die Erblichkeit der tumorösen Form der Rinderleukose. *Zeitschr. f. Tierzucht u. Züchtungsbiol.* **65**, 243–252

Burmester, B. R. and Waters, N. F. (1955). Variation in the presence of the virus of visceral lymphomatosis in the eggs of the same hens. *Poultry Sci.* **35**; 939–944

Campo, M. S., Moar, M. H., Jarrett, W. F. H. and Laird, H. M. (1980). A new papillomavirus associated with alimentary cancer in cattle. *Nature* **286**, 180–183

Cotchin, E. (1952). Neoplasia in cats. *Proc. roy. Soc. Med.* **45**, 671–674

Cotchin, E. (1957). Neoplasia in the cat. *Vet. Rec.* **69**, 425–434

Dmochowski, L., Taylor, H. G., Grey, C. E., Dreyer, D. A., Sykes, J. A., Langford, P. L., Rogers, T., Shullenberger, C. C. and Howe, C. D. (1965). Viruses and mycoplasma (PPLO) in human leukaemia. *Cancer* **18**, 1345–1368

Dutcher, R. M., Larkin, E. P. and Marshak, R. R. (1964). Virus-like particles in cow's milk from a herd with a high incidence of lymphosarcoma. *J. nat. Cancer Inst.* **33**, 1055–1064

Ferrer, J. F. (1972). Antigenic comparison of bovine type C virus with murine and feline leukaemia viruses. *Cancer Res.* **32**, 1871–1877

Ferrer, J. F., Stock, N. D. and Lin, P. (1971). Detection of replicating C-type viruses in continuous cell cultures established from cows with leukaemia. *J. nat. Cancer Inst.* **47**, 613–621

Fischinger, P. J., Schäffer, W. and Seifert, E. (1972). Murine leukaemia virus antigens in virus particles derived from 3T3 cells transformed by Murine Sarcoma Virus. *Virology* **47**, 229–235

Freeman, A. E., Gilden, R. V., Vernon, M. L. *et al.* (1974). 5-Bromo-2′-deoxyuridine potentiation of transformation of rat-embryo cells by 3-methylcholanthrene induction of Rat Leukaemia Virus gs antigen in transformed cells. *Proc. nat. Acad. Sci. USA* **70**, 2425–2419

Friend, C. (1957). Cell free transmission in adult Swiss mice of a disease having the character of a leukaemia. *J. exp. Med.* **109**, 217–228

Gardner, M. B., Arnstein, P., Rongey, R. W., Estes, J. D., Sarma, P. S., Rickard, C. F. and Huebner, R. J. (1970). Experimental transmission of feline fibrosarcoma to cats and dogs. *Nature* **226**, 807–809

Gardner, M. B., Officer, J. E., Rongey, R. W., Estes, J. D., Turner, H. C. and Huebner, R. J. (1971). C-type tumour virus genome expression in wild house mice. *Nature* **232**, 617–620

Gilden, R. V., Toni, R., Hanson, M., Bova, D., Charman, M. P. and Oroszlan, S. (1974). Immunochemical studies of the major internal poypeptide of woolly monkey and gibbon ape type C viruses. *J. Immunol.* **112**, 1250–1254

Gross, L. (1951). Pathogenic properties and 'vertical' transmission of the mouse leukaemia agent. *Proc. Soc. exp. Biol. med.* **78**, 342–348

Gross, L. (1958). Attempt to recover filterable agent from X-ray induced leukaemia. *Acta Haemat.* **19**, 353–361

Gross, L. (1961). Viral etiology of mouse leukaemia. *In*: 'Advances in Cancer Research', Vol. 6, pp. 149–180. New York: Academic Press

Gross, L. (1970). 'Oncogenic Viruses', 2nd edn, 991 pp. Oxford: Pergamon Press

Hardy, W. D. jr. Old, L. I., Hess, P. W., Essex, M. and Cotter, S. M. (1973). Horizontal transmission of feline leukaemia virus. *Nature* **244**, 266–269

Hartley, J. W., Rowe, W. P. and Huebner, R. J. (1970). Host range restrictions of Murine Leukaemia Viruses in mouse embryo cell cultures. *J. Virol.* **5**, 222–225

Hoflund, S., Thorell, B. and Winqvist, G. (1963). Experimental transmission of bovine leukosis. (abstract) 'Proc. int. Symp. on Comparative Leukaemia Research'. Doc. III, 2. Hannover, Germany

Howard, E. B., Clark, W. J. and Hackett, P. L. (1968). Experimental myeloproliferative or lymphoproliferative disease of swine. *In* (Bendixen, A. J., ed.) 'Leukaemia in Animals and Man'. Proc. 3rd Int. Symp. on Comp. Lenk. Res., pp. 255–262. *Bibl. Haemat.* no. 30. Basel: Karger.

Hsiung, G. D. (1972). Activation of guinea pig C-type virus in cultured spleen cells by 5-bromo-2′-deoxyuridine. *J. nat. Cancer Inst.* **49**, 661–662

Jarrett, W. F. H. (1966). Experimental studies of feline and bovine leukaemia. *Proc. roy. Soc. Med.* **59**, 661–662

Jarrett, W. F. H., Crawford, E. M., Martin, W. B. and Davie, F. (1964a). Leukaemia in the cat. A virus-like particle associated with leukaemia (lymphosarcoma). *Nature* **202**, 567–568

Jarrett, W. F. H., Martin, W. B., Crighton, G. W., Dalton, R. G. and Stewart, M. F. (1964b). Leukaemia in the cat. Transmission experiments with leukaemia (lymphosarcoma). *Nature* **202**, 567–568

Kalter, S. S., Heberling, R. L. and Ratner, J. J. (1973). EBV antibody in monkeys and apes. *In* (Dutcher, R. M. and Chieco-Bianchi, L., eds) 'Unifying Concepts of Leukaemia'. Proc. 5th Int. Symp. on Comp. Leuk. Res. *Bibl. Haemat.* no. 39. Basel: Karger

Kawakami, T. G., Theilen, G. H., Dungworth, G. L., Beall, S. G. and Munn, R. J. (1967). C-type particles in plasma of feline leukaemia. *Science* **158**, 1049–1050

Kawakami, T. G., Huff, S. D., Buckley, P. M., Dungworth, D. L., Snyder, S. P. and Gilden, R. V. (1972). C-type virus associated with gibbon lymphosarcoma. *Nature (New Biol.)* **235**, 170–171

Kelloff, G., Huebner, R. J., Oroszlan, S., Toni, R. and Gilden, R. V. (1970). Immunological identity of the group specific antigen of hamster-specific C-type viruses and an indigenous hamster virus. *J. gen. Virol.* **9**, 27–33

Klement, V., Nicholson, M. O., Gilden, R. V. and Huebner, R. J. (1971). Rescue of the genome of focus-forming virus from rat non-productive lines by 5-bromodeoxyuridine. *Nature (New Biol.)* **234**, 12–14

Klement, V., Nicolson, M. A., Gilden, R. V. *et al.* (1973). Rat C-type virus induced in rat sarcoma cells by 5-bromodeoxyuridine. *Nature (New Biol.)* **238**, 234–237

Laird, H. M., Jarrett, H. O., Crighton, G. W. and Jarrett, W. F. H. (1968a). An electron microscope study of virus particles in spontaneous leukaemia of the cat. *J. nat. Cancer Inst.* **41**, 867–878

Laird, H. M., Jarrett, O., Crighton, G. W., Jarrett, W. F. H. and Hay, D. (1968b). Replication of leukemogenic-type virus in cats inoculated with feline sarcoma extracts. *J. nat. Cancer Inst.* **41**, 879–893

Levy, J. A. and Pincus, T. (1970). Demonstration of biological activity of a murine leukaemia virus of New Zealand black mice. *Science* **170**, 326–327

Lévy, J. P., Boiron, M., Hollman, K. H., Haguenau, F., Thomas, M. and Friedmann, J. C. (1963). Étude au microscopique électronique par coloration négative du virus de la papillomatose bovine. *J. Microscopie* **2**, 175–182

Lieberman, M. and Kaplan, H. S. (1959). Leukemogenic activity of filtrates from radiation-induced lymphoid tumors of mice. *Science* **130**, 387–388

Livingston, D. M. and Todaro, G. J. (1973). Endogenous type-C virus from a cat cell clone with properties distinct from previously described feline type-C virus. *Virology* **53**, 142–151

Magalhães, O. (1920). Verruga dos bovideos. *Brasil-Medico* **34**, 430–431

Marshak, R. R., Hare, W. C. D., Dodd., D. C., McFeely, R. A., Martin, J. E. and Dutcher, R. M. (1967). Transplantation of lymphosarcoma in calves. *Cancer Res.* **27**, 498–504

McAllister, R. M., Nicholson, M., Gardner, M. B., Rongey, R. W., Rasheed, S., Sarma, P. S., Huebner, R. J., Hatanaka, M., Oroszlan, S., Gilden, R. V., Kabidting, A. and Vernon, L. (1972). C-type virus released from human rhabdomyosarcoma cells. *Nature* (*New Biol.*) **244**, 54–56

Miller, J. M., Miller, L. D., Olson, C. and Giletta, K. G. (1969). Virus-like particles in phytohaemagglutinin-stimulated lymphocyte cultures with reference to bovine lymphosarcoma. *J. nat. Cancer Inst.* **43**, 1297–1305

Miller, L. D., Miller, J. M. and Olson, C. (1972). Inoculation of calves with the C-type virus associated with bovine lymphosarcoma. *J. nat. Cancer Inst.* **48**, 423–428

Moloney, J. B. (1960). Biological studies on a lymphoid leukaemia virus extracted from sarcoma S. 37. I Origin and introductory investigation. *J. nat. Cancer Inst.* **24**, 933–951

Moloney, J. B. (1975). 'The Virus Cancer Program'. US Dept. Health Education and Welfare, Bethesda, Md.

Moloney, J. B. (1976). 'The Virus Cancer Program'. *Ibid.*

Moennig, V., Frank, H., Hunnsman, G., Schneider, I. and Schäffer, W. (1974). 'Properties of a mouse leukaemia virus. VII The major viral glycoprotein of Friend leukaemia virus. Isolation and physiochemical properties'. *Virology* **61**, 100–111

Nadel, E., Banfield, W., Burstein, S. and Tousimis, A. J. (1967). Virus particles associated with strain 2 guinea pig leukaemia (L2C/N-B). *J. nat. Cancer Inst.* **38**, 979–981

Olson, H. (1961). Studien über das Auftreten und die Verbreitung der Rinderleukose in Schweden. *Acta vet. Scand. Suppl. 2,* **2**, 13–46

Olson, C., Miller, L. D., Miller, J. M. and Hoss, H. E. (1972). Transmission of lymphosarcoma from cattle to sheep. *J. nat. Cancer Inst.* **49**, 1463–1467

Olson, C., Hoss, H. E., Miller, J. M. and Baumgartener, C. E. (1973). Evidence of bovine C-type (leukaemia) virus in dairy cattle. *J. Am. vet. med. Assoc.* **163**, 355–357

Parks, W. P., Livingston, D. M., Todaro, G. J., Benveniste, R. E. and Scolnick, E. M. (1973). Radioimmunoassay of mammalian type-C viral proteins. *J. exp. Med.* **137**, 622–635

Priori, E. S., Dmochowski, L., Myers, B. and Wilbur, J. R. (1971). Constant production of type-C viral particles in a continuous tissue culture derived from pleural effusion cells of a lymphoma patient. *Nature (New Biol.)* **232**, 61–62

Rauscher, F. J. (1962). A virus induced disease of mice characterized by erythrocytopoiesis and lymphoid leukaemia. *J. nat. Cancer Inst.* **29**, 515–543

Rickard, C. G., Barr, L. M., Noronha, F., Dougherty, E. 3rd. and Post, J. E. (1967). C-type virus particles in spontaneous lymphocytic leukaemia in a cat. *Cornell Vet.* **57**, 302–307

Rickard, C. G., Gillespie, J. H., Lee, K. M., Noronha, F., Post, J. E. and Savage, E. L. (1968). 'Transmission and electron microscopy of lymphocytic leukaemia in the cat'. Proc. 3rd Int. Symp. on Comp. Leuk. Res. pp. 282–284. *Bibl. Haemat.* no. 30. Basel: Karger

Rickard, C. G., Post, J. E., Noronha, F. and Barr, L. M. (1969). A transmissible virus-induced lymphocytic leukaemia in the cat. *J. nat. Cancer Inst.* **42**, 987–1014

Rubin, H., Cornelius, A. and Fanshier, L. (1961). The pattern of congenital transmission of an avian leukosis virus. *Proc. nat. Acad. Sci. USA* **47**, 1058–1069

Sarma, P. S., Log, T. and Huebner, R. J. (1970). Feline leukaemia virus detection *in vitro*. *Proc. nat. Acad. Sci. USA* **65**, 81–87

Sarma, P. S., Jain, D. K. and Hill, P. R. (1973). *In vitro* host range of feline leukaemia virus. *In* (Ito, Y. and Dutcher, R. M., eds) 'Comparative Leukaemia Research', pp. 437–440. Nagoya/Ise-Shima, University of Tokyo Press

Sherr, C. J. and Todaro, G. J. (1974). Type C viral antigens in man: I Antigen related to endogenous primate virus in human tumours. *Proc. nat. Acad. Sci. USA* **71**, 4703–4707

Snyder, S. P. and Theilen, G. H. (1969). Transmissible feline fibrosarcoma. *Nature* **21**, 1074–1075

Stenback, W. A., van Hoosier, G. L. and Trentin, J. J. (1968). Biophysical, biological and cytochemical features of a virus associated with transplantable hamster tumours. *J. Virol.* **2**, 115–1121

Stephenson, J. R., Aaronson, S. A., Arnstein, P., Huebner, R. J. and Tronick, S. R. (1974). Demonstration of two immunologically distinct xenotropic RNA viruses of mouse cells. *Virology* **61**, 411–419

Stock, N. D. and Ferrer, J. F. (1972). Replicating C-type virus in phyto-glutinin-treated buffy coat cultures of bovine origin. *J. nat. Cancer Inst.* **48**, 985–986

Theilen, H., Kawakami, T. G., Rush, J. G. and Munn, J. (1969). Repliction of cat leukaemia virus in cell suspension cultures. *Nature* **222**, 589–590

Theilen, H., Gould, D., Fowler, M. and Dungworth, D. L. (1971). C-type virus in tumor tissue of a woolly monkey (*Lagothrix sp.*) with fibrosarcoma. *J. nat. Cancer Inst.* **47**, 881–889

Ting, R. C. (1968). Biological and serological properties of viral particles from a non-producer rat. *J. Virol.* **2**, 865–868

Todaro, G. J., Arnstein, P., Parks, W. P. *et al.* (1973). A type-C virus in human rhabdomyosarcoma cells after inoculation into NIH Swiss mice treated with anti-thymocyte serum. *Proc. nat. Acad. Sci. USA* **70**, 859–862

Todaro, G. J., Sherr, C. J., Benveniste, R. E., Lieber, M. M. and Melnick, J. L. (1974). Type C viruses from baboons: isolation from normal cell cultures. *Cell* **2**, 55–61

Weinstein, R. S. and Moloney, W. C. (1965). Virus particles associated with chloroleukaemia in the rat. *Proc. Soc. exp. Biol. Med.* **118**, 459–461
Weiss, R. A., Mason, W. S. and Vogt, P. K. (1973). Genetic recombinants and heterozygotes derived from endogenous and exogenous avian RNA tumour viruses. *Virology* **52**, 535–552
WHO (1973). Immunity to cancer. *Bull. Wld. Hlth. Org.* **49**, 81–91
Wolfe, L. G. Deinhardt, F., Theilen, G. H., Rabin, H., Kawakami, T. and Bustad, L. K. (1971). Induction of tumours in marmoset monkeys with simian sarcoma virus type I (Lagothrix), a preliminary report. *J. nat. Cancer Inst.* **47**, 1115–1120
Ziegel, R. F. and Clark, H. F. (1969). Electron microscopic observations on a C-type virus in cell cultures derived from a tumor-bearing viper. *J. nat. Cancer. Inst.* **43**, 1097–1102

7
Neoplastic Diseases of Non-human Primates

I General

Until quite recent times, tumours of primates other than man were believed to be of little importance. Not only were they of infrequent occurrence, but most primates were somewhat refractory to tumour induction by carcinogenic chemicals and radiation, often with a prolonged latent period of six to seven years or more. For these reasons primates were not favoured for experimental studies. However, certain events occurred in rapid succession, which moved them to the forefront as amongst the most important animals for the study of certain forms of cancer. These events may be listed here, and studied later: 1. the discovery that a virus known as simian virus 40 (SV40), which regularly appeared in cultures of monkey kidney cells, caused cancer in baby hamsters, 2. the discovery that a virtually non-pathogenic herpesvirus of squirrel monkeys (*Saimiri sciureus*) causes rapidly fatal cancers of the lymphoma/leukaemia type in related New World species, 3. the discovery that some strains of Rous Sarcoma Virus (RSV) can cause sarcomas in some species of Old World Monkeys 4. the isolation of a D-type oncornavirus from a mammary tumour of a rhesus monkey, the Mason–Pfizer Mammary Virus (MPMV), which causes mammary tumours on sub-inoculation into other monkeys, 5. the isolation of the gibbon ape leukaemia virus (GALV), which causes leukaemia epidemics in gibbon colonies, and 6. the discovery that C-type oncornaviruses and DNA are regularly present in the germ cells and placentas of baboons.

An excellent account of 'Spontaneous and induced neoplasms in non-human primates' is given by O'Gara and Adamson (1972) in Volume I of this author's edited *Pathology of Simian Primates,* Fiennes (1972). Up to that time, only some 200 spontaneous tumours of these animals had been described in the literature, in spite of the large numbers of monkeys and apes that had been kept in zoos and research establishments over a long period. These tumours are described and classified by the authors as: 1. skeletal system; 2. sex organs 3. skin and subcutaneous tissues 4. liver and biliary system 5. pancreas and lungs 6. leukaemia 7. alimentary tract, and 8. kidney and adrenal glands. Those of most interest in the present context, as most likely to be of infectious origin, are the leukaemias and the mammary tumours.

Table 1 shows the leukaemias listed by O'Gara and Adamson. The reference to Burkitt's Lymphoma in a white-handed gibbon is intriguing and should be consulted in the original, together with the other papers referenced, by those interested. The breast tumours are listed in Table 2. No attempts seem to have been made in these cases to attempt transmission either by transplant or by filtered material. In Volume II of *Pathology of Simian Primates* will be found a short chapter on oncogenesis by Lapin and Yakovleva (1972). In this an account is given of those viruses, which have been isolated from monkey tissues and found to have oncogenic properties. These include 7 adenoviruses and one papovavirus (SV40). Included also is an account of the work of Melendez and his colleagues on *Herpesvirus saimiri* and of their own seemingly successful attempts to transmit human leukaemias by means of filtered material to macaque monkey species and baboons. These are described below.

The breakthrough relating to the oncogenic properties of *H. saimiri* came about, as so often in science, fortuitously. A virus designated herpesvirus-T had been described as causing fatal systemic disease in owl monkeys, tamarins and marmosets by Holmes *et al.* (1964) and Melnick *et al.* (1964). At the New England Primate Center, there were two spontaneous outbreaks of herpesvirus-T induced disease in owl monkeys (*Aotus trivirgatus*) the source of which was a mystery. Although there was no direct contact, it was suspected that the source of infection might be squirrel monkeys kept in the same house. As a result, as described by Hunt and Melendez (1966), the squirrel monkeys were screened for the presence of the virus by inoculation of throat and anal swab material on to rabbit kidney tissue cultures. By this means, herpesvirus-T was isolated from the squirrel monkeys, and it was shown that they were the natural hosts of the virus. It was during further studies of this virus that Melendez *et al.* (1969a) isolated another herpesvirus, of which the squirrel monkey is also a natural unaffected host; this virus was found to cause malignant lymphoma or acute lymphocytic leukaemia in owl monkeys, capuchins and marmosets (Melendez *et al.*, 1969b; Hunt *et al.*, 1970). In response to a request, Dr Hunt with great courtesy sent us slides of affected tissues and we were able to confirm, with ample authority, that the tissues were indeed lymphomatous.

Subsequent to their success with the squirrel monkey virus, Melendez *et al.* (1972a, b) and Hunt *et al.* (1972) isolated another herpesvirus with oncogenic properties from spider monkeys (*Ateles geoffroyi*); this they termed *Herpesvirus ateles*. Spider monkeys were shown to be the natural unaffected hosts of the virus, which induces malignant lymphoma in cotton-topped marmosets and tamarins.

Table 1 Spontaneous leukaemias and lymphomas in primates. (Adapted from O'Gara and Adamson, 1972)

Location	Type of tumour	Species	Sex	Age	Remarks	Reference
Blood. Enlarged spleen and lymph nodes	Chronic lymphocytic leukaemia	*Cynocephalus sphinx*			Av. WBC 36 000 with 82% lymphocytes and monocytes	Massaglia (1923)
	Chronic lymphocytic leukaemia	Green monkey				Corson-White (1929)
Lymph nodes	Lympho-sarcoma	Chacma baboon	M			Noback (1934)
Ribs, pleura, mediastinum, lymph nodes and spleen	Multiple myeloma or myelogenous leukaemia*	Mottled fox monkey		Adult		Oshima (1937)
	Lymphocytic leukaemia	Stairs monkey *(Cercopithecus albogularis stairsi)*		5½ + years		Hamerton (1942)
	Hodgkin's Sarcoma	Baboon		Unknown		Kelly (1948)
Abdominal viscera and lymph nodes	Reticulo-sarcoma	*Papio ursinus*	F	6 years +	Sub-total thyroidectomised	Gillman and Gilbert (1954)
	Leukaemia (Hodgkin)	Green monkeys			3 cases	Lapin and Yakovleva (1963)
	Malignant lymphoma	*M. mulatta*	F	7 years		Kent and Pickering (1958)
Lower lip	Lympho-sarcoma	*M. mulatta*				Gorlin. Quoted in Cohen and Goldman, (1960)
Liver, kidneys, spleen and lymph nodes	Malignant lymphoma	Gibbon *(Hylobates* sp.)	M			Newberne and Robinson (1960)
Mesentery, liver, spleen and lymph nodes	Burkitt's Lymphoma	White-handed gibbon *(Hylobates lar)*	F	4½ years		DiGiacomo (1967)
Generalised	Acute lymphocytic leukaemia	White-cheeked gibbon *(Hylobates concolor)*	M	1½ years	Massive cerebral involvement	DePaoli and Garner (1968)

*Can be translated from Japanese as either myeloma or myeloid leukaemia.

Table 2 Breast tumours of primates. (Adapted from O'Gara and Adamson, 1972)

Type of tumour	Species	Sex	Age	Remarks	Reference
Fibro-sarcoma	Green monkey				O'Conner (1947)
Carcinoma	*Macaca mulatta*	F	Older adult		Ruch (1959)
Adenocarcinoma	*Macaca mulatta*	F	10 years	Also had renal adenoma and endometrial cyst of ovary	Vadova and Gel'shtein (1956)
Adenocarcinoma	*Tupaia glis*	F		Pregnant	Elliot and Elliot (1966)
Carcinoma	*Pongo pygmaeus* (Orang-utan)	F	15 years		Brack (1966)
Carcinoma	*M. mulatta*	F	9 years	X-irradiated	Chapman (1968)

The discovery of an oncogenic herpesvirus of primates caused a great deal of excitement, because it was the first mammalian herpesvirus to be the proven cause of a cancer. It was important, too, because a human herpesvirus, the Epstein–Barr Virus (EBV) had been isolated by Epstein and Barr (1964) from a lymphoid cancer, which is common in certain parts of Africa and is known as Burkitt's Lymphoma; this will be discussed further in the next chapter. EBV, as shown by Henle *et al.* (1968) and Diehl *et al.* (1968) is the known cause of infectious mononucleosis (glandular fever) and the suspected cause of another cancer common in eastern countries, nasopharyngeal carcinoma. EBV was shown by Shope *et al.* (1973), by Falk *et al.* (1974) and Deinhardt *et al.* (1975) to induce lymphomas in marmosets and by Epstein *et al.* (1973) in owl monkeys. In spite of the universal association of EBV with two human cancers, proof that it caused them was still lacking. The sure knowledge that a mammalian herpesvirus, (*H. saimiri*) was undoubtedly the cause of lymphomas and leukaemias was a matter of some importance.

Prior to the demonstration of *H. saimiri* as a cause of neoplasia in monkeys, it had been shown that one of the oncornaviruses, the Rous Sarcoma Virus (RSV) was pathogenic for month-old rhesus and pigtail macaques, hamadryas baboons and green monkeys (Munroe and Windle, 1963; Zilber, 1964; Zilber *et al.*, 1965). The month-old monkeys developed typical tumours of Rous Sarcoma, which usually regressed. However, foetal or newborn monkeys developed fatal neoplastic tumours with metastases. These authors do not appear to have considered the 'helper' virus that would be required for positive results, and this must surely have been a 'silent' endogenous leukaemia virus. The possible existence of such in cultured human cells, which support the growth of RSV, has been mentioned earlier. RSV was also found to cause fatal cancers in marmosets of any age by Deinhardt *et al.* (1970) and by Zilber (1965). Marmosets were also shown to be susceptible to feline leukaemia virus by Deinhardt *et al.* (1970).

These studies showed that non-human primates could be of especial importance to the etiology of human cancers, because they are phylogenetically closer to man than any other animals. Useful models were already available from previous studies in birds and other mammal species to direct the course of research, and certain other viruses of the papova-and adenovirus groups, of which monkeys were the natural hosts, were known to be oncogenic in other mammals. Some 'silent' human viruses also have this property. Research material from naturally occurring tumours in non-human primates has been scanty. Susceptible and non-susceptible strains of monkeys have not been artificially bred, as with chickens and mice; as we have seen, such was a prerequisite for success in these animals.

Most of the non-human primates are almost as far removed from man in point of evolutionary time as, say, cows or seals. However, because they are derived from the same evolutionary stem, they have resemblances in respect of anatomy, physiology, response to infectious and parasitic diseases, immunology, reproductive processes, and the latent viruses they carry. Any results obtained from medical researches can, therefore, be viewed with greater confidence than those resulting from researches on rats, mice, guinea pigs and other animals. Results from researches with New World monkeys must be viewed with less confidence than those from Old World, since they are at a greater distance from the human line, and even physiologically bear less resemblance. The Great Apes are far closer to man than the monkeys, being classed with man in the super-family Hominoidea, of which there are three families: 1. Hylobatidae, with two genera – *Hylobates* (gibbons) and *Symphalangus* (siamangs), 2. Pongidae with three genera, *Pongo* (orang-utans), *Pan* (chimpanzees), and *Gorilla* (gorilla), and 3. Hominidae with a single genus *Homo,* of which there is one surviving species *Homo sapiens* and a disputable number of extinct species. A number of physiological determinants, such as the chemical nature of the blood proteins, show that the Pongidae are man's closest relatives, in the animal kingdom, with the chimpanzees in first place and the gorillas second; the gibbons are rather distant cousins; the chimpanzees are closest.

II Simian cancers caused by herpesviruses

It is evident that most vertebrates harbour one or more herpesviruses, which in normal circumstances do little or no damage. Young animals become infected in infancy and may show mild transitory symptoms, after which the infection is carried in latent form for the duration of life. In some animals, there may be more than one strain of herpesvirus, indeed several. We have encountered one such virus in frogs, which causes the Lucké Renal Carcinoma, and another in chickens, the cause of Marek's Disease. We have not, however, found herpesviruses prominently associated with cancers of cats, cattle or other animals. The ubiquitous nature of the herpes group of viruses is shown in Fig. 1. In primates, including man, the position is altered and herpesviruses become important as the causal agents of disease, and possibly of some cancers. Nahmias (1974) lists those herpesviruses which, on present evidence are or may be associated with cancers, see Table 3.

In Table 3, man figures prominently, and this group of viruses is plainly of great importance to the viral cancer picture. There are at least five herpesviruses, of which man is the natural or acquired host: *Herpesvirus simplex* type 1, *Herpesvirus simplex* type 2, varicella (chicken pox), EBV (Epstein–Barr Virus), and cytomegalovirus. Some herpesviruses carry an 'oncogene',

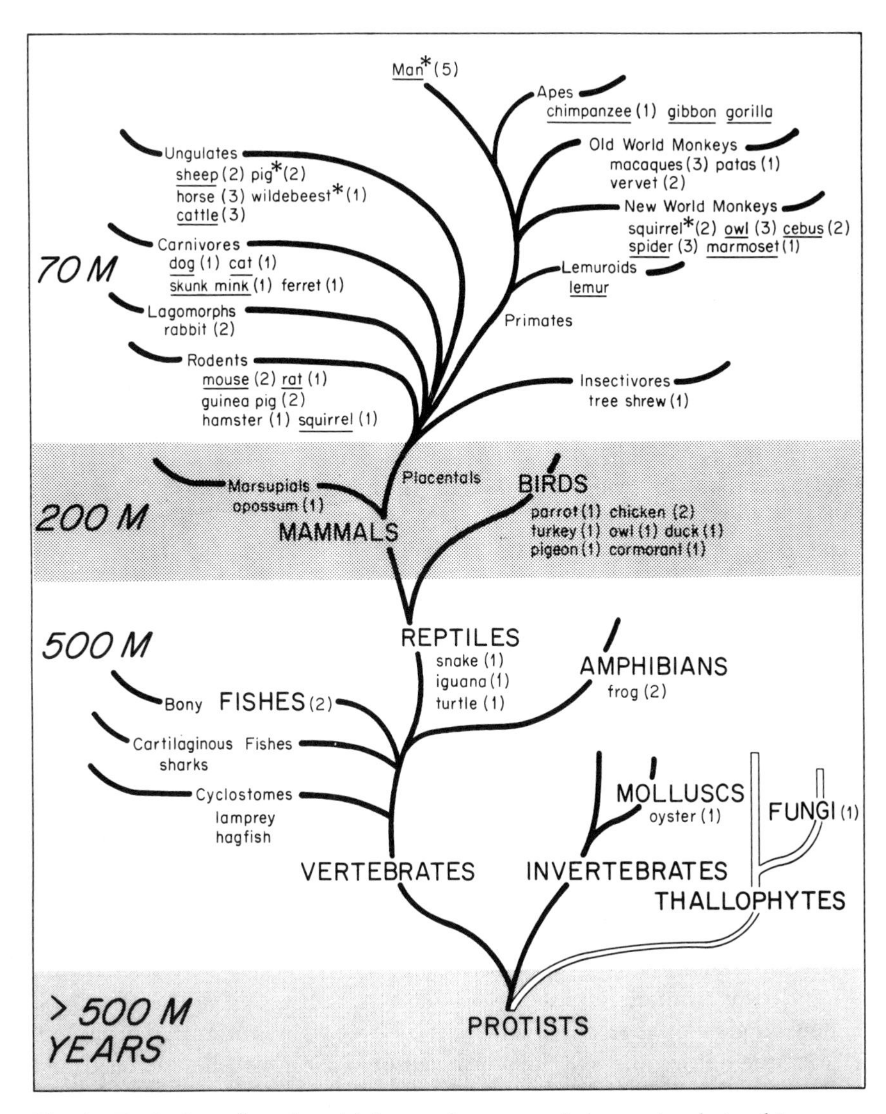

Fig. 1 Evolution of species with herpesviruses; no phylogenetic relationship among the various herpesviruses themselves is implied in this figure. Number in parentheses is the number of herpesviruses identified in each species. Species underlined are the species susceptible to herpesviruses from other species. Asterisk denotes species with herpesvirus which have been found to infect other species under natural conditions. (After Nahmias, 1974)

Table 3 Herpesviruses associated with cancer. (After Nahmias, 1974)

Herpesvirus	Species	Type of malignancy
Marek's Disease*	Chicken	Lymphoma
*H. saimiri**	Squirrel monkey	Lymphoma, leukaemia
*H. ateles**	Spider monkey	Lymphoma
Lucké Virus*	Frog	Adenocarcinoma of kidney
EB Virus*	Man	Lymphoma
H. simplex 2†	Man	Cervical cancer
EB Virus†	Man	Naso-pharyngeal carcinoma
Cottontail virus†	Cottontail rabbit	Lymphoma
Guinea pig virus†	Guinea pig	Leukaemia
H. simplex 2‡	Man	Prostatic and other urogenital cancers
H. simplex 1‡	Man	Oral (lip) and laryngeal cancers
Sheep virus‡	Sheep	Pulmonary adenomatosis of lung
Hamster virus‡	Hamster	Adenocarcinoma of bowel
Canine h. virus‡	Dog	Lymphoma
Equine h. virus 3‡	Horse	Lymphoma (in hamsters)
Bovine h. virus‡	Cow	Sarcoma & lympho-sarcomas (in hamsters)

*Strong evidence. †Good evidence. ‡Weak evidence.

capable of causing cell 'transformation' and thus possibly cancer of the host cells, others apparently do not. Herpesviruses do not normally cause serious disease in the natural host, but their effects in alien hosts may be serious.

A review of the simian herpesviruses was made by McCarthy and Tosolini (1975); more selective reviews have been made by Kalter and Heberling (1971, 1972) and Hull (1968, 1973). McCarthy and Tosolini (1975) list 37 known herpesvirus strains in primates, including man. These have been isolated from Old and New World monkeys, gorillas and chimpanzees. These authors also give a table showing the types of disease associated with the different primate herpesviruses. The types of disease are classified as: 1. neurological or generalised, in which cell-free virus is produced in culture, 2. exanthematous disease, in which cultured virus is cell-associated and possibly 'defective', 3. CMV-type, i.e. cytomegalovirus, in which virus is recoverable from 'healthy' animal tissues, 4. benign or malignant lymphoproliferative disease, with no free virus in the malignant tissue, and 5. no reported disease, but virus recoverable from healthy animal tissues. In group 4, in which we are interested, only three viruses are listed, EBV, *H. saimiri* and *H. ateles*, though, as we see from Table 3 other human viruses are suspect. The two simian viruses are from New World monkeys and only oncogenic in heterologous hosts.

Herpesvirus saimiri was isolated by Melendez *et al.* (1970) from kidney cell cultures of squirrel monkeys (*Saimiri sciureus*) during the course of investigations into another squirrel monkey virus herpesvirus-T. The squirrel monkey is the natural unaffected host of the virus and, as shown by Melendez *et al.* (1972b), 70–75% of these monkeys carry antibodies to the virus in the blood serum. Falk *et al.* (1972) found that 30% of monkeys around one year old carried antibodies, but at two years old antibodies were present in 100%. The virus is unrelated serologically to most other herpesviruses, including: *H. ateles, H. simplex,* herpesvirus-T, the virus of infectious bovine rhinotracheitis, sand rat nuclear inclusion agent, *H. suis,* ground squirrel agent, spider monkey *H. virus, H. aotus* 1 and 2, *H. saguinus,* EBV, Marek's Disease, turkey virus, and *H. sylvilagus* (Melendez *et al.,* 1972b). It is, therefore, a somewhat unusual virus.

Although no disease is caused by *H. saimiri* in the natural host, lymphomas or leukaemias result from infection in owl monkeys (*Aotus trivirgatus*), *Cebus albifrons,* cotton-topped marmosets *(Saguinus oedipus),* white-lipped marmosets (*S. nigricollis* and *S. fuscicollis*), spider monkeys (*Ateles geoffroyi*) and New Zealand white rabbits (*Oryctolagus cuniculus*) – (Melendez *et al.,* 1969b, 1970; Daniel *et al.,* 1970; Hunt *et al.* 1972; Wolfe *et al.,* 1971; Ablashi *et al.,* 1971). The incidence of cancers in marmosets and owl monkeys injected with viral filtrates approaches 100%, but the course of the disease is very variable from 4 to 116 days in marmosets and 13 to 539 days in owl monkeys. The reasons for this are unknown. Attempts to infect Old World monkeys with this virus have been unsuccessful with rhesus (*Macaca mulatta*), baboons (*Papio hamadryas*), stumptail monkeys (*M. arctoides*) and chimpanzees (*Pan troglodytes*). However, in a single experiment, Melendez *et al.* (1972b) reported the deaths from malignant disease of 2 of 6 green monkeys (*Cercopithecus aethiops*). The cells transformed by *H. saimiri* are the T lymphocytes, in contrast to those affected by EBV, which are the B lymphocytes. Solid tumours do not normally occur, but most organs and tissues are infiltrated by malignant cells. In owl monkeys and marmosets, the blood is invaded in the majority of cases, so that the condition is a true leukaemic leukaemia. Melendez *et al.* (1974) prevented the development of lymphoma in cotton-topped marmosets by injections of *H. saimiri* antiserum.

There are a great many simian herpesviruses, and no doubt many more still to be discovered; few possess oncogenic potential. As a research tool, the oncogenic viruses are very important. In primate medicine, they have a certain importance too, since infection can be transmitted horizontally from squirrel or spider monkeys to other susceptible species; they must, therefore, be kept apart. The situation with regard to the simian oncornaviruses is more complex and must now be studied.

III Simian cancers caused by oncornaviruses

1 *Endogenous oncornaviruses*

Oncornaviruses, as shown in the last chapter can for any species be separated into two classes, the 'endogenous' and the 'exogenous'. The endogenous are the host's fellow travellers, that have followed the evolutionary trail with him for millions of years and only harm him under exceptional circumstances. Most endogenous oncorna C viruses grow poorly in cells of the natural host in tissue culture, and to demonstrate their presence it is necessary to co-cultivate host cells with 'permissive' cells of a heterologous host, in which mature virus will be more readily produced. The endogenous virus is, therefore, 'xenotropic' rather than 'ecotropic'. Todaro and Huebner (1972) proposed their 'virogene hypothesis', which stated that there was genetic transfer of RNA viruses in the cells of all animals; therefore, mature virus would rarely be found in the natural host. However, Dalton *et al.* (1974) supposed that mature virus might more readily appear under certain developmental conditions, as in the developing foetus and placenta. This proposal led to the search for oncornavirus particles in the cells of reproduction, embryos, placentas and foetuses of various groups of mammals.

Mature virus particles were indeed found budding in the cytoplasm of baboon placenta cells and immature viral particles in the ova of baboons by Kalter *et al.* (1973a, 1975a, b, c, d) and Panigel *et al.* (1975). They were also found in New and Old World primates, and material from man, chimpanzee, rhesus, cynomolgus, patas and marmoset by Kalter *et al.* (1973b, 1975a, b, c); Schidlovsky and Ahmed (1973); Seman *et al.* (1975) and Vernon *et al.* (1974). The particles are more scarce in human placentas, but are nevertheless regularly present. The vertical transmission of these viruses is proven by the presence of these particles in ova, pre-implantation embryos and placentas throughout gestation. That they are xenotropic was proved by several investigators by co-cultivation with alien cells including non-human primate and human (Kalter and Heberling, 1974; Kalter *et al.*, 1975b; Benveniste, 1974b; Goldberg *et al.*, 1974; Heberling *et al.*, 1976a). As a result of intra-uterine exposure of the virus, antigenic responses are poorly developed in the natural host. After birth, these viruses disappear in their mature state and must be sought by co-cultivation techniques. They can, however, be regular, demonstrated by these means. Heberling and Kalter (1978) give the following host range of susceptible cells for their endogenous type C baboon virus: 1. *Human,* normal foetal and adult neoplastic cells, 2. *Ape,* Chimpanzee foetal and adult cells, 3. *Old World monkey,* foetal rhesus and adult African green monkey cells, and 4. *Others,* foetal and adult cells from dogs, mink, bats and horses. The virus would not replicate in cells from New World monkeys, baboons, rabbits, cats, mice or rats.

The nucleic acid of the baboon virus has been shown by Benveniste and Todaro (1974a, b) to have varying degrees of homology with that of oncornaviruses from Old World African and Asian monkeys and apes. The greatest degree of homology is with the oncornaviruses of African monkeys and least with those of man, thus reflecting the degree of evolutionary closeness of the species and supporting the claim of Todaro (1975) that there has been a stability of genetic viral information over millions of years. The stability of the genetic information during these immensely long evolutionary periods is shown further by the researches of Hellman *et al.* (1974), which shows that the baboon virus shares some common antigens with the feline endogenous oncornavirus, to which attention has been drawn in an earlier chapter.

The baboon virus has failed to induce neoplastic tumours in any primates tested or in dogs, but these animals all produce antibodies against it; baboons do not produce antibody, because of the immunological tolerance acquired during foetal life. However, the baboon virus becomes oncogenic, when hybridised with murine sarcoma virus (MSV); in this state it induces fibrosarcomas in chimpanzees, cynomolgus monkeys and marmosets. This observation is important, because one solution of the viral cancer problem could be that endogenous viruses acquire oncogenic genes from exogenous viruses infecting the host by horizontal transmission.

Kalter *et al.* (1975c) also isolated a B-type oncornavirus resembling the mouse mammary tumour virus from the prostate glands of several adult male baboons. This observation is of potential importance to the human mammary carcinoma problem. Mammary carcinoma has so far been proved of viral origin solely in mice; all attempts to trace an infectious cause in other animals have so far met with failure. This is probably because of the hormone dependence of the virus, which makes it difficult to grow in tissue culture. Strangely enough, the evidence for a viral origin of mammary cancers, except for mice, is far stronger in man than in any other animal. The presence, therefore, of a potential mammary carcinoma virus in baboon prostates is, to say the least, intriguing, having regard to the known venereal transmission of the murine virus.

D-type oncornaviruses have been isolated from New World monkeys by Kalter *et al.* (1974, 1975a, b), Seman *et al.* (1975) and Heberling *et al.* (1976b, c). The cell host range for these viruses in tissue culture includes human, chimpanzee, rhesus and canine cells. The ability of viruses to infect cells in tissues culture does not, of course, imply that they could also infect the living animal. It does, however, indicate that they are potentially dangerous, and investigators are rightly cautious in handling viruses that can infect human cells.

The endogenous oncornaviruses of non-human primates have not been implicated in the causation of any naturally occurring neoplastic process so far. They are, however, potentially of importance, because their oncogenic properties might be activated in certain circumstances, and because, being xenotropic, they could possibly come to infect human beings by horizontal transmission, causing cancers on their own account or by hybridising with endogenous human viruses. We may now study the exogenous oncornaviruses of non-human primates, a subject ably reviewed by Rabin (1978).

2 *Exogenous oncornaviruses*

Studies on the induction of tumours in non-human primates by exogenous oncornaviruses, starting with those of Munroe and Windle (1963) with Rous Sarcoma Virus, have been extensively reviewed by Rabin and Cooper (1971) and by Wolfe and Deinhardt (1972). Three oncornaviruses of the C group regularly produce malignant growths, Rous Sarcoma Virus (RSV), feline sarcoma virus (FeSV) and simian sarcoma virus (SSV). The test animals have been white-lipped and cotton-topped marmosets (*Saguinus nigricollis, S. fuscicollis* and *S. oedipus*), squirrel monkeys (*Saimiri sciureus*), rhesus monkeys (*Macaca mulatta),* grivets and vervets (*Cercopithecus aethiops* and *C. pygerythrus*) and sacred baboons (*Papio hamadryas*). The tumours induced were mostly fibro-sarcomas. Table 4, from Rabin (1978) shows the results of experiments undertaken so far.

The simian sarcoma virus (SSV) is one that we have not previously encountered. A single isolation of the virus was made from a 3-year-old pet woolly monkey (*Lagothrix* sp.) by Theilen *et al.* (1971). The animal suffered from multiple tumours, which proved to be fibro-sarcomas involving the mesentery, colon and other organs. Immature type C virus particles were demonstrated by electron microscopy in tumour tissues and in the bone marrow. Budding particles were seen also in some cells. Tumour filtrates were shown by Wolfe *et al.* (1971) to induce tumours in white-lipped marmosets.

Table 4 Tumour induction by three types of oncornaviruses in non-human primates. (After Rabin, 1978)

Primate species	RSV	FeSV	SSV
Marmosets	4+	4+	2+
Squirrel monkeys	2+	3+	Neg.
Rhesus monkeys	1–2+	1+	
Grivets and vervets	1+		
Baboons	1+		

4+ = High frequency of tumour induction, progressive tumour growth, metastases and fatalities.
1+ = Moderate degree of tumour induction and high frequency of tumour regression.

In tissue culture, the virus 'transformed' human, marmoset, dog and mouse cells. As with other sarcoma viruses, SSV was found to be deficient and a 'helper' virus was also isolated from preparations, which were active; this was termed simian sarcoma associated virus (SSAV) – (Scolnick and Parks, 1973; Wolfe *et al.,* 1972). It was also shown by Scolnick *et al.* (1972) that SSAV could also act as 'helper' to murine sarcoma virus.

Another oncornavirus the gibbon ape leukaemia virus (GALV) has been isolated from a number of gibbons suffering from lymphocytic and myelogenous leukaemias, also from normal gibbon brains. The virus was first isolated by Kawakami *et al.* (1972) from one of a pair of 3½-year-old gibbons with lymphoblastic lympho-sarcoma. Snyder *et al.* (1973) demonstrated by electron microscopy the presence of abundant type C virus particles in tumour tissue. Wong-Staal *et al.* (1976) isolated a similar virus from the spleen of another gibbon with lymphocytic leukaemia. Further isolations were made at the SEATO gibbon colony, Thailand, as reported by Kawakami and Buckley (1974). In this colony, as reported by Johnson *et al.* (1971) and DePaoli *et al.* (1973) a number of cases of both lymphocytic and granulocytic leukaemia had occurred. The isolation of 3 more strains of a similar virus from the brains of normal gibbons was reported by Todaro *et al.* (1975). Though very similar to GALV, these viruses have been designated Gibbon Brain (GBr) 1-2-3. Todaro *et al.* (1975) produced evidence that GALV-SEATO induced myelogenous leukaemia on injection into healthy gibbons.

These observations have been amply reinforced by serological studies of Kawakami *et al.* (1973) and Charman *et al.* (1975). Where leukaemia is absent from a colony of gibbons, antibodies are absent or present only in a very few animals. In colonies, in which infection is known to occur, antibodies are present in a high percentage of the gibbons. Antibody is present in all gibbons that have been artificially infected.

There is thus abundant evidence that leukaemic diseases, which occur quite commonly in gibbons, are directly caused by type C oncornaviruses, and that it is horizontally transmitted from gibbon to gibbon. Since these animals are phylogenetically not too distant from man, the discovery is of importance. It would be helpful, if the same could be demonstrated in chimpanzees or gorillas, but leukaemic infections in these animals are not of such common occurrence. Meanwhile, an interesting group of viruses has been isolated from rhesus monkeys, both healthy and suffering from extensive adenocarcinomas. The virus responsible is known as the Mason–Pfizer Virus (M-PMV).

The first isolation of M-PMV was from a female rhesus monkey with widespread adenocarcinoma of both breasts and further tumours in the left axilla, left flank, left rib cage, right ovary, adrenals, pancreas, kidneys, liver, stomach and lymph nodes. She had been received at the Mason Research Institute in January 1965, when about four years old. In April 1968, she suffered what appeared to be a prolonged oestrus with loss of weight. The weight then recovered, but the oestrus remained. In February 1969, tumours were first noted and biopsies were performed and repeated in April and May. She was then killed and post mortem examination was made. The history of the case was described by Mason *et al.* (1972). Meanwhile a virus had been isolated from the tumour tissue by Chopra and Mason (1970) and by Chopra *et al.* (1971). The virus presented some unusual features, which were described by Dalton *et al.* (1974). Some features resembled those of C-type oncornaviruses and some of B-type; it is today classified as D-type virus. The virus infected a number of different primate cells in culture and some of them were 'transformed', showing properties of malignancy. However, as shown by Fine *et al.* (1972), injection of virus filtrate into monkeys failed to induce the development of tumours, although proliferating virus was observed in biopsied tissues.

Ahmed *et al.* (1974) recovered virus similar to M-PMV from biopsied tissues of the mammary glands of 12 normal lactating rhesus monkeys. Similar virus has been recovered from cultured human cells by Gelderblom and Schwartz (1976) and Parks *et al.* (1973). This virus does not show antigenic relationships with the type C oncornaviruses of primates, though the latter show such relationships with each other and with the oncornaviruses of cats and other animals. The type C viruses of non-human primates fall into 4 groups: SSV/SSAV, GALV 1, GALV-SEATO, and GBr (gibbon brain) 1, 2, 3. Similar, if not identical, viruses have been isolated from human cells, both neoplastic and non-neoplastic. Non-human primates have also proved to be susceptible to tumour induction by type C viruses from non-primates, and as such have been extensively used experimentally. We shall, in the next chapter, consider the strange case of the Sukhumi baboons which, as described by Lapin *et al.* (1975), succumbed to an epidemic of leukaemia after being injected with large doses of filtrates derived from human cases of leukaemia.

IV Simian virus 40 (SV40)

The papovavirus was designated by Melnick (1962) to include the papilloma, polyoma and SV40 viruses, which show sufficient similarity to justify their inclusion in a single group. The only virus of importance in primate medicine is that known as SV40 or simian virus 40. During work on poliovirus a great many endogenous simian viruses were found to be contaminating monkey kidney cell cultures. As these were isolated and identified, they were assigned serial numbers as simian virus 1, 2, 3 etc; no. 40 is the papovavirus SV40. At first, the virus was believed to be of little significance, and no steps were taken to eliminate it from poliovirus and adenovirus vaccines prepared for human use. Eddy *et al.* (1961) observed that monkey kidney cell extracts induced malignant tumours when injected into newborn hamsters, and (1962a, b) showed that the agent responsible was SV40; they showed further (1962a) that sera prepared against SV40 protected the hamsters from this effect. Girardi *et al.* (1962) also induced tumours in newborn hamsters by the use of SV40 filtrates obtained from tissue cultures, thus confirming Eddy's results. The susceptibility of the hamsters diminished markedly as they increased in age. No other animals developed cancers as a result of infection with SV40, but both monkeys and human beings developed antibodies to the virus when injected with it; human cells in tissue culture readily acquired infection and were 'transformed' by the virus.

SV40 had been administered in larger or smaller doses to a great many human beings, more than a million, in polio-or adenovirus vaccines. SV40 is, moreover, usually resistant to the effects of formaldehyde and heat used in preparation of dead vaccines, so that even in these live SV40 would still be present. Once it became known that SV40 contained an oncogenic potential, there was extreme concern for the health of recipients of the vaccines, some of whom had been newborn or very young. A number of persons, who had received the virus were routinely observed over a number of years, but none has developed any cancer attributable to the virus. Today, polio-and adenovirus vaccines are only made from the kidney cells of monkeys known to be free from infection with the virus; alternatively, the virus is killed by means now known to be effective. This is simply done by a method reported by Hiatt *et al.* (1962), namely to expose the virus to visible light in the presence of toluidine blue.

V The adenoviruses

Trentin *et al.* (1962), working at Baylor University in Houston, astonished the cancer-oriented scientific community by reporting that human adenovirus no. 12 was the cause of sarcomas, when filtrates were injected into newborn hamsters. This announcement aroused a great deal of scepticism also, and it was suggested that Trentin's filtrates were contaminated either with a polyoma virus, or else with SV40. The reason for the scepticism was simple; one accepted characteristic of the adenoviruses was their inability to infect any laboratory animals. This group of viruses had only been discovered in the 1950s by Rowe *et al.* (1953), but their importance had come to be recognised. Rowe and his colleagues termed them the 'adenoid degeneration agent' (AD), because they were commonly present in the adenoids of young children with respiratory infections. A further isolation of one of these viruses was made by Hilleman and Werner (1954) from throat washings of a young military recruit; this virus was named by Hilleman *et al.* (1955) 'respiratory illness agent' (RI). Serological studies on these viruses, as reported by Huebner *et al.* (1954), showed that there were some 18 distinct serological types, and today no less than 31 are recognised. The adenoviruses are associated with respiratory illnesses and conjunctivitis and can be readily isolated from adenoids, tonsils and conjunctival secretions. Hence they were given a third name 'adenoidal-pharyngeal-conjunctival' (APC) group. How ever, as reported by Enders *et al.* (1956) agreement was reached at a meeting in New York City to adopt the term 'adenovirus group' for them.

Infective particles of adenoviruses were demonstrated under the electron microscope by a number of investigators, including Hilleman *et al.* (1955). Some of the virus types are normally latent and cause no overt disease symptoms. Others are associated with a variety of febrile illnesses, accompanied by pharyngitis or conjunctivitis, and other respiratory ailments. There are three groups of these viruses based on the symptoms they cause; 1. types 3, 4, 7 and 14 occur in epidemics of acute respiratory infections and are rarely if ever latent, 2. type 8 causes epidemics of a specific type of kerato-conjunctivitis, and 3. types 1, 2, 5 and 6 are often latent in lymphoid tissues; from time to time, they may be associated with sporadic cases of respiratory disease. The other types are normally latent and not commonly associated with respiratory diseases.

In spite of the scepticism aroused by his results, Trentin and his colleagues had been deliberately testing the adenoviruses for their oncogenic potential because of their morphological similarity to other viruses, which possessed it, and because of the well-known tendency of some viruses to cause febrile diseases in one species and cancer in another. The types tested were 2, 3, 7, 9, 10, 11, 12 and 14. Of these, only type 12 caused neoplastic changes, which Trentin was easily able to prove were not due to contamination of his material with other viruses. As with so many neoplasms, recovery of virus from tumour tissue is only accomplished with great difficulty, but this was achieved by Connor and Martin (1966). Trentin's results were confirmed by Huebner *et al.* (1962). The matter was thus placed beyond dispute, and they were able to demonstrate that adenovirus type 18 also induced neoplasms in newborn hamsters. They also showed that hamsters could be protected against both types 12 and 18 by immune sera prepared against them. Other adenovirus groups were subsequently found to possess oncogenic potential for newborn hamsters, but were variable in their effect. In summary, types 12 and 18 possess high oncogenic potential; others, types 1, 7, 8, 14, 21 and 24 possess low oncogenic potential. Those with low oncogenic potential not only induce tumours in a smaller percentage of hamsters treated, but have also a longer incubation period. The tumours induced are mostly sarcomas, but those associated with types 1, 7, 8, 14 and 24 are usually lymphomas, though they may prove to be lymphosarcomas.

Subsequent to the work on human adenoviruses, studies were made of those affecting or carried by non-human primates. Hull *et al.* (1965) showed that 5 of 7 adenoviruses isolated from rhesus and cynomolgus monkeys induced tumours in newborn hamsters. A virus with very high oncogenic potential was also isolated from an African green monkey (*Cercopithecus aethiops*). This virus was 100% effective in causing tumours in hamsters, and induced them also in 3 of 21 newborn rats and 1 of 5 newborn mice. Oncogenic potential has also been demonstrated in a chicken adenovirus by Sarma *et al.* (1965) and in a bovine adenovirus by Darbyshire (1966). For a review of the simian adenoviruses, see Heberling (1972).

The adenoviruses lead us logically from the infectious tumours of non-human primates to those of man to be studied in the next chapter.

References

Ablashi, D. V., Loeb, W. F., Valerio, M. O., Adamson, R. H., Armstrong, G. R., Bennet D. C. and Heine, V. (1971). Malignant lymphoma with lymphocytic leukaemia induced in Owl Monkeys by *Herpesvirus saimiri*. *J. nat. Cancer Inst.* **47**, 837–855

Ahmed, M., Korol, W., Yeh, I., Schidlovsky, G. and Mayyasi, J. A. (1974). Detection of Mason-Pfizer Monkey Virus infection using human KC cells carrying Rous Sarcoma Virus genome. *J. nat. Cancer Inst.* **53**, 383–387

Benveniste, R. E. and Todaro, G. J. (1974a). Multiple divergent copies of endogenous type-C virogenes in mammalian cells. *Nature* **252**, 170–173

Benveniste, R. E. and Todaro, G. J. (1974b). Evolution of C-type viral genes. Inheritance of exogenously acquired viral genes. *Nature* **252**, 456–459

Brack, M. (1966). Carcinoma solidum simplex Mammae bei einem Orang-utan. (*Pongo pygmaeus*). *Zbl. allg. Path. path. Anat.* **109**, 474–480

Chapman, W. L. (1968). Neoplasia in non-human primates. *J. Am. vet. med. Ass.* **153**, 872–878

Charman, H. P., Kim, N., White, M., Marquaret, H., Gilden, R. V. and Kawakami, T. G. (1975). Naturally and experimentally induced antibodies to defined mammalian type-C viral proteins in primates. *J. nat. Cancer Inst.* **55**, 1419–1424

Chopra, H. C. and Mason, M. M. (1970). A new virus in a spontaneous mammary tumour of a rhesus monkey. *Cancer Res.* **301**, 2081–2086

Chopra, H. C., Zelljadt, I., Jensen, E. M., Mason, M. M. and Woodside, N. J. (1971). Infectivity of cell cultures by a virus isolated from a mammary carcinoma of a rhesus monkey. *J. nat. Cancer Inst.* **46**, 127–137

Cohen, D. N. and Goldman, H. M. (1960). Oral disease in primates. *Ann. NY Acad. Sci.* **5**, 889–909. (*Vide* ref. to Gorlin)

Conor, J. D. and Martin, A. (1966). Isolation of adenoviruses from tissue cultures of adenovirus type 12 induced hamster tumours. (abstract). *Proc. Am. Assoc. Cancer Res.* **7**, 14

Corson-White, E. P. (1929). Chronic lymphatic leukaemia in a Green Monkey. *Arch. Path.* **81**, 1019

Dalton, A. J., Melnick, J. L., Bauer, H., Beaudreau, G., Bentzvelen, P., Bolognesi, D., Gallo, R., Graffi, A., Haguenau, F., Heston, W., Huebner, R., Todaro, G. and Heine, V. I. (1974). Retroviridae. *Intervirol.* **4**, 201–206

Daniel, M. D., Melendez, L. V., Hunt, R. D., King, N. W. and Williamson, M. E. (1970). Malignant lymphoma induced in rabbits by *Herpesvirus saimiri* strains. *Bact. Rev.* **271**, 197

Darbyshire, J. H. (1966). Oncogenicity of bovine adenovirus type 3 in hamsters. *Nature* **211**, 102

Deinhardt, F., Wolfe, L. G., Theilen, G. H. and Snyder, S. P. (1970). ST-feline fibrosarcoma virus: induction of tumours in marmoset monkeys. *Science* **167**, 881

Deinhardt, F., Falk, L., Wolfe, L. G., Pacega, J., and Johnson, D. (1975). Responses of marmosets to experimental infection with Epstein-Barr-Virus (EBV). *In* (Thé, G., Epstein, M. A. and Zur Hausen, H., eds). 'Oncogenesis and Herpesvirus II'. IARC Sci. Pub. No. 11, Int. Ag. for Res. in Cancer, Lyon, France

DePaoli, A. and Garner, F. M. (1968). Acute lymphocytic leukaemia in a white-cheeked gibbon (*Hylobates concolor*). *Cancer Res.* **28**, 2559–2561

DePaoli, A., Johnson, D. O. and Noll, W. W. (1973). Granulocytic leukaemia in white-handed gibbons. *J. Am. vet. med. Ass.* **163**, 624–628

Diehl, V., Henle, G., Henle, W. and Kohn, G. (1968). Demonstration of a herpes group virus in cultures of peripheral leucocytes from patients with infectious mononucleosis. *J. Virol.* **2**, 663–669

DiGiacomo, R. F. (1967). Burkitt's lymphoma syndrome in white-handed gibbon (*Hylobates lar*). *Cancer Res.* **27**, 1178–1179

Eddy, B. F., Borman, G. S., Berkeley, W. H. and Young, R. D. (1961). Tumours induced in hamsters by injection of rhesus monkey kidney cell extracts. *Proc. Soc. exp. Biol. Med.* **107**, 191–197

Eddy, B. E. Borman, G. S., Grubbs, G. E. and Young, R. D. (1962a). Identification of the oncogenic substance in rhesus monkey kidney cell cultures as Simian Virus 40. *Virology* **17**, 65–75

Eddy, B. E. Grubbs, G. E. and Young, R. D. (1962b). Persistent infection of human carcinoma and primary chick embryo cell cultures with Simian Virus 40. *Proc. Soc. exp. Biol. Med.* **111**, 718–722

Elliot, O. S. and Elliot, M. W. (1966). Breast cancer in a tree shrew (*Tupaia glis*). *Nature* **211**, 1105

Enders, J. F., Bell, J. A., Dingle, J. N. Francis, T. jr., Hilleman, M. R., Huebner, R. J. and Payne, A. M-M. (1956). 'Adenoviruses'; Group name proposed for new respiratory-tract viruses. *Science* **124**, 119–120

Epstein, M. A. and Barr, Y. M. (1964). Cultivation in vitro of human lymphoblasts from Burkitt's malignant lymphoma. *Lancet* **1**, 252–253

Epstein, M. A., Hunt, R. D. and Rabin, H. (1973). Pilot experiments with EB Virus in owl monkeys (*Aotus trivirgatus*): I Reticuloproliferative disease in an inoculated animal. *Int. J. Cancer* **12**, 309–318

Falk, L. A., Wolfe, L. G. and Deinhardt, F. (1972). Isolation of *Herpesvirus saimiri* from blood of Squirrel Monkeys (*Saimiri sciureus*). *J. nat. Cancer Inst.* **48**, 523–530

Falk, L. A., Wolfe, L. G. and Deinhardt, F. (1974). Epstein-Barr Virus (EBV): Experimental injection and lymphoma induction in marmoset monkeys. *Fed. Proc.* **33**, 739

Fiennes, R. N. T-W-. (1972). 'Pathology of Simian Primates', Vols I and II. Basel: Karger

Fine, D. L., Kingsbury, E. W., Valerio, M. G., Kubicek, M. T., Landon, J. C. and Chopra, H. C. (1972). Simian tumour virus proliferation in inoculated *Macaca mulatta. Nature, (New Biol.)* **238**, 191–192

Gelderblom, H. and Schwartz, H. (1976). Relationship between the Mason-Pfizer Virus and HeLa Virus: Immuno-electroscopy. *J. nat. Cancer Inst.* **56**, 635–637

Gillman, A. and Gilbert, C. (1954). Some connective tissue diseases (amyloidosis, arthritis, reticulo-sarcoma) in the baboon (*Papio ursinus*). *S. Afr. J. med. Sci.* **19**, 112

Girardi, A. J., Sweet, B. H., Slotnick, V. B. and Hilleman, M. R. (1962). Development of tumours in hamsters inoculated in the neonatal period with vacuolating virus SV40. *Proc. Soc. exp. Biol.* Med. **109**, 649–660

Goldberg, R. J., Scolnick, E. M. Parks, W. P., Yakovleva, L. A. and Lapin, B. A. (1974). Isolation of a primate type-C virus from a lymphomatous baboon. *Nat. J. Cancer* **14**, 722–730

Hamerton, A. E. (1942). Primary degeneration of the spinal cord in monkeys, a study in comparative pathology. *Brain* **65**, 193–204

Heberling, R. L. (1972). The simian adenoviruses. *In* (Fiennes, R.N.T-W-, ed.) 'Pathology of Simian Primates', Vol. II, pp. 573–591. Basel: Karger

Heberling, R. L. and Kalter, S. S. (1978). Endogenous RNA oncornaviruses of non-human primates. *In* (Chivers, D. J., Ford, E. N. R., eds) 'Recent Advances in Primatology', Vol. IV. London: Academic Press

Heberling, R. L., Kalter, S. S., Barker, S. T. and Weislow, O. S. (1976a). Isolation and biological properties of endogenous baboon (*Papio cynocephalus*) type C viruses. *Bibl. Haematol.* **431**, 158–160

Heberling, R. L. Barker, S. T., Helmke, R. J., Smith, G. C. and Kalter, S. S. (1976b). 'Isolation of a retrovirus from Squirrel Monkey (*Saimiri sciureus*) lung cell culture', p. 216 (569). *Abs. Ann. Mtg. Am. Soc. Microbiol.*

Heberling, R. L., Barker, S. T., Kalter, S. S., Smith, G. C. and Helmke, R. J. (1976c) Oncornavirus: Squirrel monkey (*Saimiri sciureus*) lung culture. *Science* **195**, 289–292

Hellman, A., Peebles, P. T., Strickland, J. E., Fowler, A. K., Kalter, S. S., Oroszlan, S. and Gilden, R. V. (1974). Baboon virus isolate M7 with properties similar to feline virus RD 114. *J. Virol.* **14**, 133–138

Henle, G., Henle, W. and Diehl, V. (1968). Relation of Burkitt's tumour-associated virus to infectious mononucleosis. *Proc. nat. Acad. Sci. USA* **59**, 94–101

Hiatt, C. W., Gerber, P. and Friedman, R. M. (1962). Photodynamic inactivation of the vacuolating virus, SV40. *Proc. Soc. exp. Biol. Med.* **109**, 230–232

Hilleman, M. R. and Werner, J. H. (1954). Recovery of new agent from patients with acute respiratory illness. *Proc. Soc. exp. Biol. Med.* **85**, 183–188

Hilleman, M. R., Tousimis, A. J. and Werner, J. H. (1955). Biophysical characterization of the RI (RI-67) viruses. *Proc. Soc. exp. Biol. Med.* **89**, 587–593

Holmes, A. W., Caldwell, R. G., Dedmon, R. I. and Deinhardt, F. (1964). Isolation and characterization of a new herpes virus. *J. Immunol.* **92**, 602–610

Huebner, R. J., Rowe, W. P., Ward, T. G., Parrott, R. H. and Bell, J. A. (1954). Adenoidal-Pharyngeal-Conjunctival agents. A newly recognized group of common viruses of the respiratory system. *New England J. Med.* **251**, 1077–1086

Huebner, R. J., Rowe, W. P. and Lane, W. T. (1962). Oncogenic effects of human adenovirus types 12 and 18. *Proc. nat. Acad. Sci. USA* **48**, 2051–2058

Hull, R. N. (1968). 'The simian viruses' Virology Monographs. Vol. 2, pp. 1–66. New York: Springer Verlag

Hull, R. N. (1973). The simian herpesviruses. *In* (Kaplan, A. S., ed.) 'The Herpesviruses', pp. 390–424. London: Academic Press

Hull, R. N., Johnson, I. S., Culbertson, C. G., Reimer, C. B. and Wright, H. F. (1965). Oncogenicity of the simian adenoviruses. *Science* **150**, 1044–1046

Hunt, R. D. and Melendez, L. V. (1966). Spontaneous Herpes-T infection in the owl monkey. (*Aotus trivirgatus*). *Path. Vet.* **3**, 1–26

Hunt, R. D., Melendez, L. V., King, N. W., Gilmore, C. E., Daniel, M. D., Williamson, M. E. and Jones, T. C. (1970). Morphology of a disease with features of malignant lymphoma in marmoset and owl monkeys inoculated wth *Herpesvirus saimiri*. *J. nat. Cancer Inst.* **44**, 447–465

Hunt, R. D., Melendez, L. V., King, N. W. and Garcia, F. G. (1972). *Herpesvirus saimiri* malignant lymphoma in spider monkeys: a new susceptible host. *J. med. Primatol.* **1**, 114–128

Johnson, D. O., Wooding, W. L., Tanticharoenyos, P. and Bolognesi, C. N. (1971). Malignant lymphoma in the gibbon. *J. Am. vet. Ass.* **159**, 563–566

Kalter, S. S. and Heberling, R. L. (1971). Comparative virology of primates. *Bact. Rev.* **35**, 310–364

Kalter, S. S. and Heberling, R. L. (1972). Serological evidence of viral infection in South American monkeys, *J. nat. Cancer Inst.* **49**, 251–259

Kalter, S. S. and Heberling, R. L. (1974). Isolation of C-type viruses from baboon (*Papio cynocephalus*) placental tissue. *Abs. Abs. Ann. Mtg. am. Soc. Microbiol.* p. 233 (VI94)

Kalter, S. S., Helmke, R. T., Heberling, R. L., Panigel, M. F., Strickland, J. C. and Hellman, A. (1973a). C-type particles in human placentas. *J. nat. Cancer Inst.* **50**, 1081–1084

Kalter, S. S., Helmke, R. J., Panigel, M., Heberling, R. L., Felsburg, P. J. and Axelrod, L. R. (1973b). Observations of apparent C-type particles in baboon (*Papio cynocephalus*) placentas. *Science* **179**, 1332–1333

Kalter, S. S., Kuntz, R. E., Heberling, R. L., Helmke, R. J. and Smith, G. C. (1974). C-type particles in a urinary bladder neoplasm induced by *Schistosoma haematobium*. *Nature* **251**, 440

Kalter, S. S., Heberling, R. L., Hellman, A., Todaro, G. J. and Panigel, M. (1975a). Viruses in the transmission of cancer. *Proc. roy. Soc. Med.* **68**, 135–140

Kalter, S. S., Heberling, R. L., Hellman, A., Todaro, G. J. and Panigel, D. M. (1975b). C-type particles in baboon placentas. *Proc. roy. Soc. Med.* **68**, 135–140

Kalter, S. S. Heberling, R. L., Smith, G. C. and Helmke, R. J. (1975c). C-type viruses in chimpanzee (*Pan* sp.) placentas. *J. nat. Cancer Inst.* **55**, 735–736

Kalter, S. S., Heberling, R. L., Helmke, R. J., Panigel, M., Smith, G. C., Kraemer, D. C., Hellman, A., Fowler, A. K. and Strickland J. E. (1975d). C-type particles in normal human placentas. *In* (Ito, Y. and Dutcher, R. M., eds) 'Comparative Leukaemia Research 1973, Leukemogenesis.' Basel: Karger

Kawakami, T. G. and Buckley, P. M. (1974). Antigenic studies on gibbon type-C viruses. *Transplant Proc.* **61**, 193–196

Kawakami, T. G., Huff, S. D., Buckley, P. M., Dungworth, D. I., Snyder, S. P. and Gilden, R. V. (1972). C-type virus associated with gibbon lymphosarcoma. *Nature (New Biol.)* **235**, 170–171

Kawakami, T. G., Buckley, P. M., McDowell, T. S. and DePaoli, A. (1973). Antibodies to simian C-type virus antigen in sera of gibbons (*Hylobates* sp.). *Nature (New Biol.)* **246**, 105–107

Kelly, A. L. (1948). Annual Report of the Hospital and Research Committee. *Zoonooz. Res.* **21**, 7

Kent, S. P. and Pickering, J. E. (1958). Neoplasma in monkeys (*Macaca mulatta*). Spontaneous and irradiation produced. *Cancer* **11**, 138–147

Lapin, B. A. and Yakovleva, L. A. (1963). 'Comparative Pathology in Monkeys.' Springfield: Thomas

Lapin, B. A. and Yakovleva, L. A. (1972). Oncogenesis of primates. *In* (Fiennes, R. N. T-W-, ed) 'Pathology of Simian Primates', Vol. II., pp. 725–747. Basel: Karger

Lapin, B. A., Yakovleva, L. A., Indzhiia, L. V., Agrba, V. Z., Tsiripova, G. S., Voevodin, A. F., Ivanov, M. T. and Djatchenko, A. G. (1975). Transmission of human leukaemia to nonhuman primates. *Proc. roy. Soc. Med.* **68**, 141–145

Mason, M. M., Bogden, A. E., Illievski, V., Esber, H. J., Baker, J. Q. and Chopra. H. C. (1972). History of a rhesus monkey adenocarcinoma containing virus particles resembling oncogenic RNA viruses. *J. nat. Cancer. Inst.* **48**, 1323–1331

Massaglia, A. C. (1923). Leukaemia in the monkey. Note on a syndrome in a monkey (*Cynocephalus sphinx*) similar to that of chronic lymphatic leukaemia in man. *Lancet* **204**, 1056–1057

McCarthy, K. and Tosolini, F. A. (1975). A review of primate herpesviruses. *Proc. roy. Soc. Med.* **68**, 145–150

Melendez, L. V., Daniel, M. D., Garcia, F. G., Fraser, C. E. O., Hunt, R. D. and King, N. W. (1969a). *Herpesvirus saimiri.* I Further characterization studies of a new virus from the squirrel monkey. *Lab. Anim. Care.* **19**, 372–377

Melendez, L. V., Hunt, R. D., Daniel, M. D., Garcia, F. G. and Fraser, C. E. O. (1969b). *Herpesvirus saimiri.* II Experimentally induced malignant lymphoma in primates. *Lab. Anim. Care* **19**, 378–386

Melendez, L. V., Hunt, R. D., Daniel, M. D., Fraser, C. E. O., Garcia, F. G. and Williamson, M. E. (1970). Lethal reticuloproliferative disease in *Cebus albifrons* monkeys by *Herpesvirus saimiri. Int. J. Cancer* **6**, 431–435

Melendez, L. V., Hunt, R. D., King, N. W., Barahona, H. H., Daniel, M. D., Fraser, C. E. O. and Garcia, F. G. (1972a). A new lymphoma virus of monkeys: *Herpesvirus ateles. Nature (New Biol.)* **235**, 182–184

Melendez, L. V., Hunt, R. D., Daniel, M. D., Fraser, C. E. O., Barahona, H. H., King, N. W. and Garcia, F. G. (1972b). *Herpesvirus saimiri* and *ateles*; their role in malignant lymphomas of monkeys. *Fed. Proc.* **31**, 1643–1650

Melendez, L. V., Hunt, R. D., Garcia, F. G., Daniel, M. D. and Fraser, C. E. O. (1974). Prevention of lymphoma in cotton-topped marmosets by *Herpesvirus saimiri* anti-serum. *J. med. Primatol.* **3**, 213–220

Melnick, J. L. (1962). Papova virus group. *Science* **135**, 1128–1130

Melnick, J. L., Midulla, M., Wimberley, I., Barrera-Ora, J. G. and Levy, B. M. (1964). A new member of the herpesvirus group isolated from South American marmosets. *J. immunol.* **92**, 596–601

Munroe, J. S. and Windle, W. F. (1963). Tumours induced in primates by chicken sarcoma virus. *Science* **140**, 1415–1416

Nahmias, A. J. (1974). The evolution (evovirology) of herpesviruses. *In* (Kurstak, E. and Maramorosch, K. eds) 'Viruses, Evolution and Cancer', New York and London: Academic Press

Newberne, J. W. and Robinson, V. B. (1960). Spontaneous tumours in primates: report of two cases with notes on the apparent low incidence of neoplasms in sub-human primates. *Amer. J. vet. Res.* **21**, 150–155

Noback, C. V. (1934). Report of the veterinary. *NY Zool. Soc. Ann. Rep.* **38**, 39–43

O'Conner, P. (1947). Occurrence of tumors in zoo animals. *Animaland* **14**, 2–4

O'Gara, R. W. and Adamson, R. H. (1972). Spontaneous and induced neoplasms in non-human primates. *In* (Fiennes, R. N. T-W-, ed.) 'Pathology of Simian Primates', Vol. I, pp. 190–238. Basel: Karger

Oshima, F. (1937). Studies on tumours in wild animals, *Gann.* **31**, 220–223

Panigel, M., Kraemer, D. C., Kalter, S. S., Smith, G. C. and Heberling, R. L. (1975). Ultrastructure of cleavage stages and preimplantation embryos of the baboon. *Anat. Embryol.* **147**, 45–62

Parks, W. P., Gilden, R. P., Bykovsky, A. F., Miller, G. G., Zhdanov, V. M., Soloviev, V. D. and Scolnick, E. M. (1973). Mason-Pfizer Virus characterization: a similar virus in a human amniotic cell line. *J. Virol.* **12**, 1540–1547

Rabin, H. (1978). Studies in non-human primates with exogenous type-C and type-D oncornaviruses. *In* (Chivers, D. J. and Ford, E. H. R., eds) 'Recent Advances in Primatology', Vol. IV. London: Academic Press

Rabin, H. and Cooper, R. W. (1971). Tumor production in squirrel monkeys (*Saimiri sciureus*) by Rous Sarcoma Virus. *Lab. Anim. Sci.* **21**, 1032–1049

Rowe, W. P., Huebner, R. J., Gilmore, L. K., Parrott, R. H. and Ward, T. G. (1953). Isolation of a cytopathogenic agent from human adenoids undergoing spontaneous degeneration in tissue culture. *proc. Soc. exp. Biol. Med.* **84**, 570–573

Ruch, T. C. (1959). 'Diseases of Laboratory Primates'. Philadelphia: Saunders

Sarma, P. S., Huebner, R. J. and Lane, W. T. (1965). Induction of tumours in hamsters with an avian adenovirus (CELO). *Science* **149**, 1108

Schidlovsky, G. and Ahmed, M. (1973). C-type virus particles in placentas and fetal tissues of rhesus monkeys. *J. nat. Cancer Inst.* **51**, 225–233

Scolnick, E. M. and Parks, W. P. (1973). Isolation and characterization of a primate sarcoma virus: mechanism of rescue. *Int. J. Cancer* **12**, 138–147

Scolnick, E. M., Parks, W. P., Todaro, G. J. and Aaronson, S. A. (1972). Immunological characterization of primate C-type virus reverse transcriptase. *Nature (New Biol.)* **235**, 35–40

Seman, G., Levy, B. M., Panigel, M. and Dmochowski, L. (1975). Type-C particles in placentas of the cottontop marmoset (*Saguinus oedipus*), *J. nat. Cancer Inst.* **54**, 251–252

Shope, T., Dechairo, D. and Miller, G. (1973). Malignant lymphoma in cottontopped marmosets following inoculation of Epstein-Barr Virus. *proc. nat. Acad. Sci, USA* **70**, 2487–2491

Snyder, J. P., Dungworth, D. L., Kawakami, T. G., Callaway, E. and Law, D. T-L. (1973). Lymphosarcoma in two gibbons (*Hylobates lar*) with associated C-type virus. *J. nat. Cancer Inst.* **5**, 89–94

Theilen, H., Gould, P., Fowler, M. and Dungworth, D. L. (1971). C-type virus in tumour tissue of a woolly monkey (*Lagothrix* sp.) with fibrosarcoma. *J. nat. Cancer Inst.* **47**, 881–889

Todaro, G. J. (1975). Evolution and modes of transmission of RNA tumour viruses. *Am. J. Pathol.* **81**, 590–605

Todaro, G. J. and Huebner, R. J. (1972). The viral oncogene hypothesis: new evidence. *Proc. nat. Acad. Sci. USA* **69**, 1009–1015

Todaro, G. J., Lieber, M. M., Benveniste, R. E., Sherr, C. J., Gibbs, C. J. jr. and Gajdusek, D. C. (1975). Infectious primate type-C viruses: three isolates belonging to a new sub-group from the brains of normal gibbons. *Virology* **67**, 335–343

Trentin, J. J., Yabe, Y. and Taylor, G. (1962). Tumor induction in hamsters by human adenovirus. (abstract) *Proc. am. Assoc. Cancer Res.* **3**, 369

Vadova, A. V. and Gel'shtein, V. T. (1956). Spontaneous tumours in catarrhine monkeys according to the date obtained in the monkey colony of the Sukhumi medico-biological section. *In* (Utkin I. A., ed.) 'Theoretical and Practical Questions of Medicine and Biology in Experiments on Monkeys', pp. 137–158. New York: Pergamon Press

Vernon, M. L. McMahon, J. M. and Hackett, J. J. (1974). Additional evidence of type-C particles in human placentas. *J. nat. Cancer Inst.* **52**, 987–989

Wolfe, L. G. and Deinhardt, F. (1972). Oncornaviruses associated with spontaneous and experimentally induced neoplasia in nonhuman primates. *In* (Goldsmith, E. I. and Moor-Jankowski, J., eds) 'Medical Primatology', Part III, pp. 176–196. Proc. 3rd. Conf. exp. Med. Surg. Primates, Lyon

Wolfe, L. G., Deinhardt, F., Theilen, H., Rabin, H., Kawakami, T. and Bustad, L. K. (1971). Induction of tumors in marmoset monkeys with simian sarcoma virus type I (*Lagothrix*): a preliminary report. *J. nat. Cancer Inst.* **47**, 1115–1120

Wolfe, L. G., Smith, R. D., Hoekstra, J., Marczynska, B., Smith, R. K., McDonald, R., Northrop, R. L. and Deinhardt, F. (1972). Oncogenicity of feline fibrosarcoma viruses in marmoset monkeys: pathogenic, virologic and immunologic findings. *J. nat. Cancer Inst.* **49**; 519–539

Wong-Staal, F., Gallo, R. C., Sarngadharan, M. G. and Gillespie, J. H. (1976). Proviral sequences in spleen of a gibbon with spontaneous lymphoma. *Proc. Am. Assoc. Cancer Res.* **17**, 120

Zilber, L. A. (1964). 'Some data on the interaction of Rous Sarcoma Virus with mammalian cells'. Int. Conf. on Avian Tumour Viruses, *Nat. Cancer Inst. Mon.* **17**, 26–71

Zilber, L. A. (1965). Pathogenicity and ocogenicity of Rous Sarcoma Virus for mammals. *Progr. exp. Tum. Res.* 7, 1–48

Zilber, L. A., Lapin, B. A. and Adzighitov, F. I. (1965). Pathogenicity of Rous Sarcoma Virus for monkeys. *Nature* **205**, 1123–1124

8

Infectious Viruses in Human Cancer

I General

In chapter 4, the question was asked as to whether all sarcomas and other tumours of chickens could be certainly attributed to infectious viral causes? The recovery of virus even from tumours induced by chemicals and radiation made it seem probable that a viral genetic input was essential for cell 'transformation' to the cancerous state. It was found that, though the evidence supported such a conclusion, it was not strong enough for it to be stated as a fact. With other animals, too, man apart, the position is similar. Could it be stated with certainty that all animal cancers, however induced, required the intervention of a virus for their development, it could be stated with a degree of certainty that the same must be true of human cancers. So long as there is a lingering doubt that factors other than viruses could upset the genetic apparatus of the cell, then the possibility remains that in man the viral factor is of lesser importance than it is in other animals. Nevertheless, there is certainty or virtual certainty that some human cancers at any rate are directly caused by viral infections, and it will be our object in this chapter to review the evidence.

Certain facts of a general nature have emerged in previous chapters, and these may be recapitulated before proceeding to a study of the human cancers, which may be associated with the known oncogenic viral groups.

1 *The carrier state*

(*a*) *The herpesvirus group*. Man in common with other animals possesses his own range of herpesviruses, of which there are 5 peculiar to his species: 1. *Herpesvirus simplex* 1, 2. *Herpesvirus simplex* 2, 3. Epstein–Barr Virus (EBV), 4. Varicella/Zoster virus (chicken pox/shingles), and 5. Cytomegalovirus. These viruses can all cause disease symptoms, generally quite mild but sometimes severe; all can exist in the body in the carrier state once active symptoms have subsided, and can infect other persons, who have not experienced them. All these viruses, except that of chicken pox are believed to possess 'oncogenes' and are suspected of being associated with human cancers. EBV, at any rate, is positively incriminated as a cause of human cancer.

(*b*) *The oncornavirus group*. As with all other animals so far studied, human cells carry in their genome the genes of oncornaviruses, which are transmitted vertically from one generation to the next, and free viral particles can be demonstrated in foetuses and placentas. Oncornaviruses are potentially oncogenic, either through possession of an oncogene or the ability to acquire one through gene exchange with another virus of the same group. Viruses of this group are, therefore: (*i*) continuously present in human tissues; (*ii*) definitely oncogenic, (*iii*) liable to acquire new properties by recombination with other viruses of the same group, such as might be acquired by contact infection, (*iv*) they can be transmitted either horizontally or vertically, and (*v*) they are frequently found in association with human neoplasms.

(*c*) *The papilloma virus group*. Human warts and papillomas are certainly caused by a virus of the papova group, of which papilloma virus is one; the virus frequently exists in the body in the 'carrier' state. Papillomas of the bladder sometimes become neoplastic, as in cattle, in which a similar virus is known to be responsible for the neoplastic condition.

(*d*) *The adenovirus group*. Man is either susceptible to or carrier of at least 31 adenoviruses, a number of which possess oncogenic potential to a greater or lesser degree.

Man is, therefore, continuously exposed to four groups of viruses, which are potentially oncogenic, and some are found in increased concentration in some tumour tissues; these same groups of viruses are known to be the direct cause of animal cancers. Human cells in tissue culture can support the growth of such viruses from animal sources, and can furthermore evidently supply 'helper' virus to deficient sarcoma viruses from chickens and cats.

2 *Difficulties of research in man*

Successes in oncornavirus research in chickens and mice depended to a large degree on the availability or development of cancer susceptible and insusceptible strains. Even so, success was not easily achieved with filtered material, until tumours had been transmitted through several generations of tissue transplants. In man, there are no very definite race susceptibilities, which could aid the investigator. Though there appear to be some familial susceptibilities, there is obviously no possibility of breeding cancer susceptible and resistant strains, and to do so would in any case take an unrealistically long time. Furthermore, experiments to transmit cancers either by transplants or filtrates are not possible. The solution of the lympho-sarcoma problem in cattle was eventually achieved by studies of high- and low-incidence herds, the existence of which favoured investigation. Even so, difficulties were encountered, because of the slow development of tumours in a potentially long-lived animal, and because some tumours regressed in the younger beasts. In an even longer-lived animal, such as man, these difficulties are further enhanced.

The problem will be more easily solved, when more is known of the biology and genetics of cancer cells. Certain properties place them in a separate category from their healthy parent cells such as: 1. they grow progressively at the expense of the tissue in which they are located, and are not subject to the 'contact inhibition' rule, which regulates and orders the growth of normal cells in a tissue, 2. they are 'immortal' in the sense that they will multiply continuously in tissue cultures, whereas normal cells fade out and die after a certain number of generations, and 3. they can grow in the absence of oxygen (anaerobic growth), whereas normal cells require oxygen for their growth, replication and maintenance of their activities. It may be argued from this, that cancer cells have reverted to a more primitive type as a result of the loss or neutralisation of some gene or genes. For example, some cells of more primitive organisms, such as yeast fungi, share these properties with vertebrate cancer cells. In this case, neutralisation or destruction of the relative genes could be equally well effected by chemicals, radiation or infective viruses.

These suppositions would carry greater conviction, were it not for the demonstration that all animal cancers studied in depth have been found, with the reservations expressed above, to be due to viral involvement. Weight is thereby lent to the argument that such fundamental changes of the cell biology are more likely to arise from the addition of new genetic information by the incorporation of viral genes, than from destruction of gene capacity. This argument is further supported by the demonstration that oncornaviruses, at any rate, possess a specific 'oncogene', which is distinct from those genes by which the virus reproduces and assembles its enzymes, and provokes the host cell into manufacturing new DNA for replication. Evidence that incorporation of a viral oncogene in the host genome was essential to cell transformation would certainly show that human cancers, like those of other animals, were dependent on the presence of a carcinogenic virus.

Further strange anomalies exist in that oncogenic properties are found in only four of the many groups of viruses and only in some members of each. Again, there is the strange anomaly that these viruses prove to be carcinogenic in some animals, but not in others. It is evident from this that more knowledge is required of the relationships of viruses with their host cells, and of the ways – immune processes, Interferon production – in which the host reacts to the presence of the viruses. When such problems are better understood, the situation relating to human cancer should become more clear. Even if direct evidence, as by transmission experiments, cannot be obtained, yet the viruses involved could be isolated and identified, and means of prophylaxis could be devised with a reasonable degree of confidence.

II Transmission of human cancers by cell transplants

Experiences of cell transplants in human beings are understandably rare, and there are no known instances in newborn infants. Such cases as have occurred are discussed by Gross (1970), who should be consulted for details and references. Cell transplants have been: 1. deliberate or accidental, 2. in the same or a different individual, and 3. where the transplant has occurred in a different individual, either in one unrelated or else in one with ties of consanguinity.

1 *Cell transplants in the same individual*

Cell transplants in the same individual are in effect what occurs when metastases arise naturally, so that there would be seemingly little sense in repeating nature's own experiments. However, it was supposed that a patient's own cancer cells might create an immune response, if they were transferred to a site that was less suitable for them, and that such a response might cause regresssion of the original tumours. Some such transfers failed to grow; some grew to a certain extent and then regressed; in other cases, a malignant growth appeared in the new site and required to be excised. No favourable effects on the original tumours resulted.

There have also been a number of instances, in which malignant tumours have been accidentally spread to new sites during the course of surgery.

2 *Cell transplants in different unrelated individuals*

Southam *et al.* (1957) Southam and Moore (1958) described how they injected cancer cells, obtained from a human source and maintained for three years in tissue culture, into human volunteers suffering from cancer. The transplants, in most cases, caused small nodules which regressed in 4–6 weeks. In 4 patients, the cancer cells reproduced malignant tumours, which in 2 grew progressively; in 1 patient, the cancer cells caused metastases in new sites and a fresh malignant condition was induced. They also injected lines of human malignant cells into 14 healthy volunteers at the Ohio State Penitentiary in Columbus. No cancers resulted, only small sterile abscesses or nodules, which appeared after 3–4 weeks; they all disappeared spontaneously, except those which were excised for microscopic studies. The response to the cancer cells was thus inflammatory, not malignant.

A small number of cases is described by Gross (1970), in which cancers have been accidentally transferred from a patient to a healthy individual. In one such, a man contracted a carcinoma of the glans penis, identical with a cervical carcinoma from which his wife was suffering. There was also the celebrated and tragic case of a French medical student described by Lecène and Lacassagne (1926) and by Katz (1930). The student, Henri Vadon, was aspirating fluid from an operation wound, following the removal of her breast from a woman suffering from an adenocarcinoma. The syringe slipped and punctured Vardon's left hand injecting some of the fluid deep into his palm. Two and a half years later, a fusiform sarcoma developed at the injection site. The tumour metastasised necessitating amputation of the arm, but further tumours developed above the collar bone and in the neck; Vadon died a year later. Thus, accidental injection of adenocarcinoma cells resulted in the development of generalised fibro-sarcoma – not carcinoma as might have been expected. There was no confirmed history of malignancy in Vadon's family on either his father's or his mother's side.

A number of cases are quoted by Gross (1970) in which cancers have developed from kidney transplants. In such cases, the donor kidneys have been taken from patients dying from malignant tumours of the lungs or other organs. The recipients have also been receiving treatment with immuno-suppressive drugs; when this has been discontinued the transplanted kidneys have in most cases been rejected and the cancers have regressed naturally; in other cases, death from generalised malignancy has occurred. Kidneys from donors with malignant disease are no longer used for transplants.

3 *Cell transplants in different, but related, individuals*

It would be expected from results with chickens and mice, that transmission of human cancers would more readily occur between blood relations than between unrelated persons. There are, however, few records of such, though the familial incidence of cancers is well established, *vide infra*. Weber *et al.* (1930) reported a case of adenocarcinoma transmitted by a mother of 27 years to her child. The child was born normally and in good health, but the mother died 3 months later of generalised melano-sarcoma in which the uterus and placenta were involved. The child died at 8 months from generalised melano-sarcoma with large tumours in the liver. The implications are that the child was infected during intra-uterine life either by malignant cells from the mother reaching its liver, highly improbable, or by a virus.

Another case is described by Balacesco and Tovaru (1936). A young mother had a small nodule in her breast, which ulcerated. In spite of this she continued to suckle her baby, which developed a tumour of the lower lip when 11 months old. Both tumours, of mother and child, were then removed surgically. The breast was an adenocarcinoma, that of the child a spindle-cell sarcoma.

The third case to be considered was reported by Scanlon *et al.* (1965). A 50-year-old woman suffered from a melanoma on the middle of her back.

The tumour was excised, but generalised metastases appeared three years later. At this point, the patient's mother, a woman of 80 years, insisted on having her daughter's tumour implanted in herself. In accordance with her wishes, a portion of the tumour was embedded in one of her muscles. The tumour grew rapidly in the muscle, and she died 86 days later of disseminated metastatic melanoma, even though the original tumour had been excised.

Although the evidence for transplantation of human tumour cells is scanty and inconclusive, it is yet of some importance as showing that under certain circumstances malignant tumours can be transplanted from one person to another. Such conditions include blood relationship and immuno-suppression. It is also significant that in the two cases, quoted, mammary carcinoma cells have induced sarcomas; this might indicate that the original tumour was induced by a pluripotent virus, rather than that the transplant arose from injected cells.

III Familial incidence of human cancers and epidemiological studies

It is fully recognised that some cancers occur more frequently in some families than in others, and appear regularly in successive generations. The question is outlined briefly, but adequately, by Gross (1970), who gives full documentation. This statement is true both of the hard tumours and of the leukaemias. We are not concerned here, therefore, as to whether this happens but as to why it happens. Several possibilities present themselves: 1. is it because of genetic instability, which gives rise to cancer susceptible and cancer resistant strains as with mice? 2. if so, does this genetic instability originate in the patients' own genes, or is it dependent on the incorporation of the genes of oncogenic viruses in the host genome? 3. if this is so, is the development of cancer directly dependent on the presence of the resident viral genes or are other factors also involved such as (*a*) exposure to co-carcinogens, chemicals, radiation, air pollution, tobacco smoke and so on, or (*b*) to horizontal infection with viruses, which hybridise with the resident viruses? 4. if the development of the cancers is not directly dependent on genetical defects, is it due to a diminished resistance to oncogenic viruses acquired by horizontal infection?

There are also racial susceptibilities to certain forms of cancer, which could be due either to genetical or environmental causes. Such have not been extensively investigated, though primary cancers of the stomach are commoner in Caucasian than in coloured people. Primary cancer of the liver is rare in all races except Bantu peoples, living in Africa; the reason may be the consumption of foods, especially groundnuts, which are contaminated with moulds. The well-known Burkitt's Lymphoma occurs only in certain geographical regions, especially in Africa, which coincide with areas of endemic malaria, a disease possibly linked with its development. Primary tumours of the naso-pharynx are common in the Far East, especially amongst the Chinese. The latter two cancers are believed to be caused by the Epstein–Barr Virus (EBV) as will be discussed under. Cancers of the lungs and bronchi are associated with industrial areas, in which the atmosphere is polluted, and are commoner in heavy smokers. Again, many cancers are associated with ways of life and occupation. Workers in plants manufacturing certain chemicals are more prone to certain cancers, especially those working with asbestos. This gives little in the way of clues as to whether these influences are the direct cause of the cancers, or merely co-carcinogens, which activate viruses. In chickens and mice, it will be recalled, viruses can be isolated from tumours induced by radiation or chemicals.

If cancers are indeed infectious and the infection is passed horizontally, this should become clear from studies of their epidemiology. This matter also was reviewed by Gross (1970), who finds no evidence of horizontal spread of any forms of human cancer, except for isolated instances of apparent small epidemics of leukaemia. Some statistical evidence has been adduced by Levy (1974), that there is a higher incidence of leukaemia amongst children in households where cats are kept. Although the leukaemias of chickens and mice are transmitted vertically, it will be recalled that those of cats are transmitted horizontally.

An analysis of the incidence of cancers in the population of the United Kingdom for the years 1972–1978 is given in Table 1. The evidence arising from this analysis belies any suggestion of an epidemic incidence. The figures, indeed, show a remarkable uniformity of the total incidence of malignant disease as a cause of death from year to year and of the percentages dying from particular cancers year by year. It is only when the incidence of particular cancers is studied as percentages of the total cancer incidence, that some variation is noted. For example, as cancers of the lungs and bronchi increase, those of other organs, such as the stomach, decrease. This might indicate some competition between the location of cancers, and suggest that any infectious agents with pluripotent properties behave as opportunists, occupying sites most readily made available by co-carcinogens.

It is probable that special factors, as with Burkitt's Lymphoma and cervical carcinoma, are responsible for the spread of certain types of cancer; yet, if of infectious origin the great majority of cancers are evidently transmitted vertically, otherwise the statistics would show greater fluctuation of cancer incidence and indicate some evidence of epidemics.

It is now necessary to consider suspect viral groups in relation to the human cancers, which they are supposed to cause. As with the non-human primates, we shall first study the herpesviruses.

Table 1 Incidence of cancer in the UK, 1972–1978. (Figures for Total population in millions, elsewhere, thousands)

	1972	1973	1974	1975	1976	1977	1978
Total population	55 781	55 913	55 922	55 901	55 881	55 852	55 835
Total deaths	674.00	669.7	667.4	662.5	681.0	655.3	651.1
Deaths % population	1.2	1.2	1.2	1.3	1.2	1.2	1.2
Total malignancies	136.1	137.3	139.2	139.9	143.2	142.9	141.9
Mal. as % population	0.24	0.25	0.25	0.25	0.26	0.26	0.25
Mal. as % deaths	20.2	20.5	20.9	21.1	21.0	21.8	21.8
Stomach	14.1	13.8	13.6	13.5	13.4	12.9	12.7
Lungs and bronchi	35.8	36.4	37.4	37.2	38.0	38.3	38.4
Breast	12.6	12.9	12.7	13.1	13.2	13.4	13.2
Uterus	4.1	4.3	4.1	4.1	4.2	4.1	4.1
Leukaemia	3.5	3.4	3.6	3.6	3.6	3.6	3.7
Stomach % deaths	2.1	2.1	2.0	2.0	2.0	2.0	1.9
do. % malignancies	10.4	10.1	9.8	9.7	9.3	9.0	8.9
Lung etc. % deaths	5.3	5.4	5.6	5.6	5.8	5.8	5.9
do. % malignancies	26.3	26.5	26.9	26.6	26.5	26.8	27.0
Breast % deaths	1.9	1.9	1.9	2.0	1.9	2.0	2.0
do. % malignancies	9.3	9.4	9.1	9.4	9.2	9.3	9.3
Uterus % deaths	0.6	0.6	0.6	0.6	0.6	0.6	0.6
do. % malignancies	3.0	3.1	3.0	2.9	2.9	2.9	2.9
Leukaemia % deaths	0.5	0.5	0.5	0.5	0.5	0.6	0.6
do. % malignancies	2.6	2.5	2.5	2.6	2.5	2.5	2.6

N.B. A great deal of publicity has in recent years been focused on the serious situation arising from cancers of the lungs and bronchi, largely attributed to the habit of smoking. It has not apparently been realised that the situation with regard to the female cancers, especially those of the breast, amounting to some 12% of all cancers in only one half the population is almost as serious. Since breast cancers and those of the cervix, in the USA the second highest in incidence of the female cancers, are as will be seen below, probably due to viral infections, it is surprising that more effort is not directed to finding means of prevention.

IV Herpesviruses and cancers of man

The herpesviruses, as we have seen, are widely distributed in the animal kingdom, each group of animals being carriers of strains peculiar to it. In spite of this, they appear always to be horizontally transmitted often shortly after birth. Of the five sub-types, which normally infect man, *Herpesvirus simplex* type 1 is perhaps that, which resembles most the prototype. Infection with this virus is almost universal, and it is passed to children by the mother shortly after birth; the child may suffer from herpetic lesions or aphthous sores of the mouth and lips for a time, but these pass and the infant then becomes a carrier of the virus for life. Sometimes skin lesions around the mouth will recur in the form of 'cold sores' at times of stress or mild febrile illness. *Herpesvirus simplex* type 2 and cytomegalovirus (CMV) are both venereally transmitted, this being the commonest way in which infection is passed. *Herpesvirus simplex* 2 causes herpetic lesions of the genitalia. CMV is normally quiescent, being located in the salivary glands and kidneys; under conditions of stress or immuno-suppression, it can, however, cause serious and even fatal disease; it has on occasion been found associated with tumours, but the evidence for its involvement with any cancers is insufficiently strong to be worth pursuing in this work. Both CMV and Epstein–Barr Virus (EBV) can also be transmitted orally, the commonest means, according to Nahmias (1974), being during the act of kissing! EBV acquired by an uninfected person from a carrier will then cause the disease of infectious mononucleosis (glandular fever), and can be isolated for long periods from the throat (Miller *et al.*, 1973). *Herpesvirus varicellae*, the virus of varicella/zoster (chicken pox/shingles) is, of course, epidemically spread, but once acquired remains latent for life in the nervous system, and may cause shingles in later life.

Characteristically, herpesviruses cause either a sub-clinical or non-lethal infection in their natural host, but severe and often fatal disease in other, even closely related, species; these infections include neoplastic conditions such as lymphomas. The sites, where the viruses are located in the natural host, outside infancy, are listed by Nahmias (1974) as being, in general, those from which the virus can most easily spread. These include: 1. the skin and oral or genital mucous membranes, 2. the upper and occasionally lower respiratory tract, as in the laryngo-tracheitis group of diseases of chickens, cats, dogs, horses, cattle and varicella/zoster of man, and 3. the eyes, as with chickens, cats, dogs, pigs and sometimes with *H. simplex* 1 and varicella/zoster. Many of these viruses can at times also infect internal organs such as the liver, and in particular lymphoid tissues, thence being found in circulating leucocytes. The affinity for lymphocytes, according to Nahmias (1974) may be associated with the means by which the viruses persist in the body, thereby enabling the virus to escape destruction by circulating antibodies. It is, however, in such sites that cell transformation usually occurs; they may well be unnatural sites, invasion of which occurs mainly in unnatural hosts or when the powers of resistance are embarrassed. Such a hypothesis could explain the odd behaviour of a virus such as EBV which causes, or so it is believed, three widely differing conditions in human hosts, depending on the persons involved and the geographical region, where the infection occurs. We shall consider first among the herpesviruses the problem of the association of EBV with human neoplasia.

1 *Epstein–Barr Virus – infectious mononucleosis*

EBV was first described by Epstein and Barr (1964) arising from their studies of Burkitt's Lymphoma. Epstein, as he informed the author at the time, was suspicious of the virus because of its resemblance to that of the Lucké Renal Adenocarcinoma. It is, however, more logical to study EBV first as the cause of infectious mononucleosis (glandular fever). This disease was not recognised as a form of leukaemia until its association with EBV had been established and extensive researches had been undertaken.

The connection between EBV and infectious mononucleosis (IM) was first suggested by Henle *et al.* (1968) and by Diehl *et al.* (1968). IM is a benign febrile disease, accompanied by lassitude and depression, whose chief significance for the patient is its long course of several months before the symptoms subside. There is, during its course, a marked increase of atypical lymphocytes in the peripheral blood, lymph nodes, spleen and bone marrow, in fact a reversible hyperplasia of the lymphatic system. As shown by Macfarlane Burnet there exist in the body two cell lines of lymphocytes, which are indistinguishable under the microscope, but can be differentiated by certain laboratory tests. The one line is known as the T cell line, because their activity is dependent on the thymus gland. The other line is that of the B cells, produced in the bone marrow and other lymphatic tissues. Both are concerned in immune reactions and antibody production, and their activities are interrelated. Lymphoblastoid (parent) T cells can be grown in tissue cultures; B cells cannot in normal circumstances. It is, however, the B cells, which become infected with EBV, and they can be readily grown in tissue cultures if EBV DNA is present in their genome. EBV DNA is never found in T cells, and it is evident that EB virus is a pathogen of B lymphocytes probably infecting them at the lymphoblastoid stage, transforming them and stimulating unnatural replication. The transformed B lymphocytes are, therefore, neoplastic and IM patients are suffering from B cell leukaemia. The question may reasonably be asked: 'Why then do IM patients recover naturally from a cancer, when other cancers are progressive and usually end fatally?' The answer is that they do not. B cell lymphocytes from IM patients, who have apparently recovered, retain EBV DNA in their genome for life, and cell lines from them can be cultured *in vitro* to the end of their days. This is one human cancer of indubitable infectious origin, in which the body's defence systems keep its progress in control and, though patients remain infected, active symptoms disappear.

It is often said by those who find it difficult to accept the possibility that human cancers could be of infectious origin, that no proof of such infection has so far been produced. Such statements are erroneous, and this must give added confidence to the proposal that other human cancers, especially those that may be associated with herpesviruses, may also be infectious.

2 *Epstein–Barr Virus – Burkitt's Lymphoma*

While the association of EBV with IM is today generally accepted, doubts are still expressed as to whether this virus is the cause of Burkitt's Lymphoma (BL). Yet, the presence of EBV DNA can be demonstrated in virtually all tumours, which are definitely identified as belonging to the Burkitt syndrome (Pagano, 1974). Pagano states the case as follows:

> The remarkable association of the Epstein–Barr virus and two forms of human malignancies, one of the African Burkitt's lymphoma and the other naso-pharyngeal carcinoma, has been demonstrated and now quantitatively defined in approaching 100% of cases. This demonstration is a stronger link to the virus than the finding of EBV antibodies that are ubiquitous in normal patients as well as those with Burkitt's lymphoma; even those with disease do not necessarily have high titers of antibodies.

He continues to stress that it would be truly amazing if only in endemic areas of Africa did EBV select tumour tissue as a 'passenger' with no causal relationship.

The story of Burkitt's Lymphoma is an intriguing one. The first case was described in 1904 by the well-known medical missionary Dr (later Sir Albert) Cook, when working in Uganda. Although many cases must have been submitted for treatment since then and the disease existed in many areas outside Uganda, it was not formerly recognised as a distinct syndrome requiring investigation. Attention was first drawn to this by Denis Burkitt (1958), then a young surgeon working at the Makerere University Medical School's Mulago Hospital, at Kampala, Uganda. The tumour was identified as a malignant lymphoma by O'Conor and Davies (1960), O'Conor (1961) and O'Conor *et al.* (1965). These tumours occur in children between the ages of 2 and 14 years, being most frequent at the age of 5; both sexes are affected in equal proportion. The tumours characteristically develop in the bones of the jaw, both mandible and maxilla, often extending to the orbit. The tumours also invade the ovaries bilaterally, the thyroid glands and the kidneys. Adrenals, heart and stomach may also become involved, and lesions sometimes develop in the spinal cord causing paralysis. Unlike other lymphomas, the lymphatic glands, spleen and peripheral blood are not affected. The condition is rapidly progressive and surgical intervention ineffective. More recently, some drugs have been found, which may ameliorate the condition or even effect cures.

Burkitt (1958, 1962a, b, c, d; Burkitt and Davies, 1961) recognised that the lymphoma bearing his name is an entity distinct from other lymphomas; he also believed that the occurrence of the syndrome was confined to definable geographical limits. If this were so, he believed that it must be transmitted by a secondary vector, probably a mosquito, and he sought to confirm his suspicions relating to its distribution. This he did by sending questionnaires to doctors and hospitals throughout central and other possible parts of Africa. The replies appeared to confirm his ideas, and he then went on a 10 000 mile journey to confirm personally what he had learned, accompanied by two other doctors, Dr Williams and Dr Nelson. Results confirmed the existence of his tumour syndrome across tropical Africa from the east coast to Dakar in Senegal in the extreme west. In East Africa, including Uganda, the tumour does not occur above an altitude of 5000 feet. The distribution of the tumour is given by Burkitt (1962b) as: 1. throughout Uganda, Kenya and Tanzania anywhere, except at altitudes of more than 5000 feet and in the islands of Zanzibar and Pemba, 2. throughout Malawi, Zambia and Zimbabwe, only in the great river valleys and on the shores of Lake Nyassa, 3. throughout the coastal regions of Mozambique, and 4. thought to be unknown in South Africa, though since Burkitt's report rare cases have been reported there and some from Zanzibar.

The belief that a mosquito-borne virus might be responsible was quickly abandoned, but the coincidence of the endemic areas with those of endemic malaria led to the supposition that chronic malaria infection might be a co-factor in tumour induction. Little is heard of this belief today, but there is evidently some merit in it. Since Burkitt's original reports on distribution, similar tumours have been reported elsewhere in the world. The majority have occurred in areas ecologically similar to the African cases. Ten Seldam *et al.* (1966) found 35 cases during six years in Papua and New Guinea, and stated that the Burkitt-like syndrome was the commonest malignancy of childhood in these areas. Khan (1964) found a case in an Arab child in Zanzibar, and Beltrán *et al.* (1966) described 6 cases in children 4–5 years old in Colombia, South America. A further case was described by Booth *et al.* (1964) in a 5½-year-old child of Lebanese parents in Sierra Leone. Some Burkitt-like tumours have also been described from outside the Burkitt ecological definition. Chapman and Jenkins (1963) described 5 cases in children in Natal, South Africa, and Gluckman (1963) 3 cases in white children in Johannesburg; these cases could all have been in malarious areas, so could still conform to the Burkitt formula. However, a number of lymphomas, resembling Burkitt's, have also been described from the United States and Canada, and one in an English 15-year-old schoolgirl. There are reasons, discussed below, for believing that the latter cases, in spite of superficial resemblances, were not of the true Burkitt syndrome.

It was established by Pulvertaft (1964), Epstein and Barr (1964, 1965), Epstein *et al.* (1967) and other workers that Burkitt's Lymphoma cells could be grown and maintained over long periods in tissue culture, in which they revealed special characteristics. They proved to be lymphoblastoid cells of the B line, which can only be grown in tissue culture if EBV DNA is present in the genome.

The search for a candidate causative virus was, as with all cancers, an arduous and prolonged process. The case for a virus being implicated was, however, taken seriously and the Imperial Cancer Research Fund of London sent a unit to join the East African Virus Research Institute at Entebbe, Uganda. The results of the investigation were reported by Simons and Ross (1965). Herpesvirus had been isolated from tumours, but proved to be *Herpesvirus simplex* 1. Since this virus is normally present in the oral mucous membranes of children and can be isolated in a percentage of healthy persons, it was concluded that the virus had no etiological significance. Woodall *et al.* (1965), working at the same Institute, conducted parallel investigations and achieved the same results. Meanwhile, Epstein and Barr (1964, 1965) studied the problem on a different basis. Whereas Simons, Ross, and Woodall, had added Burkitt tumour fragments to cultures of hamster kidney and human foetal lung cells, Epstein and his colleagues grew the actual tumour cells in culture. The cells grew well and were examined by electron microscopy. By this means, they found in the cells virus particles resembling herpesvirus and having a characteristic morphology. Epstein concluded that the virus present was not *Herpesvirus simplex*, because it was some 20% smaller. This conclusion was subsequently confirmed by Epstein and Barr (1964), Epstein *et al.* (1967) and by Henle and Henle (1966) and Henle *et al.* (1968) on the basis of biological and immunological data. Epstein *et al.* (1967) found similar viral particles in a New Guinea case of the syndrome. Epstein's viral particles are, however, only to be found in cultured cells; they are not be seen in cells from primary tumours. Other viruses and *Mycoplasma* have also been isolated from these tumours, but are not believed to be concerned with their development, and one may, therefore, ignore them.

Epstein–Barr Virus has today been established as a herpesvirus in its own right distinct from *Herpesvirus simplex*. Its association with Burkitt's Lymphoma has been reviewed by Epstein (1970) and by Klein (1971, 1972). A general review of the herpesviruses, including EBV is contained in the Academic Press publication *The Herpesviruses* (Kaplan, 1973). EBV has since those days been extensively studied and its oncogenic potential fully established; for example Gerber and Hoyer (1971) have shown that EBV can stimulate the induction of thost cell DNA synthesis. Whereas electron microscopy and immuno-fluorescence studies fail to demonstrate the presence of viral particles in primary tumour tissue, EBV viral DNA has been demonstrated in such tissues, though it is absent from the tissues of healthy persons (Pagano, 1974).

The evidence for EBV as the cause of Burkitt's Lymphoma is so strong that it cannot seriously be doubted. We may now turn to the association of EBV with naso-pharyngeal carcinoma.

3 *Epstein–Barr Virus – naso-pharyngeal carcinoma*

Naso-pharyngeal carcinoma has a wider geographical distribution than Burkitt's Lymphoma, although it is found mostly in the Far East and parts of Africa, and affects mostly persons of Asiatic races. Being a carcinoma, the malignant cells are of epithelial origin, though tumours are heavily infiltrated with lymphocytes. Nevertheless, as shown by Henle *et al.* (1970b) and de Thé (1972), the tumour cells contain large amounts of EBV DNA; this is discussed by Pagano (1974). EBV has also been associated with Hodgkin's Disease by Levine *et al.* (1971) and with Guilain–Barré Syndrome by Grose and Feorino (1972). A great many detailed studies of EBV have been made since the references quoted. These are documented and to a certain extent discussed in the annual reports of the Virus Cancer Program (1975–1978). This virus is certainly the etiological cause of one human cancer, infectious mononucleosis, and its etiological role in the causation of Burkitt's lymphoma and naso-pharyngeal carcinoma is widely accepted. It would appear that the production of prophylactic preparations to protect populations at risk cannot be long delayed. It is now necessary to study the possible role of the herpes simplex group of viruses in causing human cancers.

4 *The role of* Herpesvirus simplex *in human cancer*

Herpesvirus simplex may reasonably be regarded as the normal resident herpesvirus of human beings. Very few people fail to acquire infection shortly after birth and they retain the virus for the duration of life. Its natural history is well described by Nahmias (1974). *Herpesvirus simplex* (HSV), like similar viruses of other animals, mostly causes insignificant disease in its natural host, but can cause diseases of great severity in other primates, including neoplasia. Even in man, the *Herpesvirus simplex* viruses can at times cause serious and even fatal diseases in the following ways: 1. by means of infection of the nervous system, 2. by causing disease of the foetus, leading to abortion, or of the newborn, 3. very occasionally by causing disease of the adult, and 4. possibly by causing neoplasia. Encephalitis or meningitis usually occurs in the very young; the virus, during its latent phase, survives in nervous tissues, and in some cases may not become well adapted to residence in such a site. Similarly, when a cancer is caused, whether in the natural or an alternate host, it is probably due to an aberration in the integration of the viral genome in cells in which the virus might persist in its latent phase (Nahmias, 1974).

It was first demonstrated by Nahmias and Dowdle (1968) that there were two types of HSV, HSV_1 being isolated from the mouth and respiratory tract and HSV_2 from the genital organs. The venereal transmission of HSV_2 was demonstrated by Centifanto *et al.* (1972) who showed that it caused genital herpetic lesions. However, a French physician, John Astuc, had published an account of genital herpes as a venereal disease as long ago as 1736 (Goodheart, 1970). This disease and cervical carcinoma, the second most frequent cancer of women, are both of more frequent occurrence in women who have first had sexual experience at a young age and are promiscous. A number of pathological studies, documented by Rapp and Koment (1974) have indicated that HSV_2 infection, associated with genital herpes, often precedes cervical dysplasia, which is considered to be a precursor of cervical carcinoma. Serological studies by Rawls *et al.* (1972, 1973) have confirmed the linkage.

It was serological studies, such as these, which gave the first indication of the association of HSV_2 with cervical carcinoma. This is of interest in contrast with the sequence of events leading to the involvement of EBV with Burkitt's Lymphoma. EBV, it will be recalled, was first identified as a result of isolation of entire viral particles from tissue cultures of Burkitt's Lymphoma cells. It was positively identified as the cause of this tumour, because viral DNA with its characteristic nucleotide sequences was found to be associated with virtually 100% of the tumours it was supposed to cause. Studies were, however, also being made on the 'early' antigens of EBV by Klein *et al.* (1966, 1967) and by Henle *et al.* (1970a). These are proteins produced in the early stages of viral synthesis, and their detection either by analysis or by serological methods is important to the search for oncogenic viruses.

Meanwhile, the association of HSV_2 with cervical carcinoma, in spite of the circumstantial evidence outlined above, remained uncertain, because viral DNA could only be identified in some 30% of tumour tissues. The presence of viral DNA in tumour tissues shows that viral genes are present in the host cell genome, but not necessarily that they are playing an active part in a neoplastic or other pathological process. Its absence does not indicate that the virus concerned is not responsible for a neoplastic condition, for two reasons. First, the DNA may be present in amounts too small to be detected by analysis; the work on chicken and mouse neoplasias showed that virus or viral genetic information might only be detected after several generations of tumour transplantation. Secondly, if the 'hit and run' hypothesis is correct, then the virus may initiate the neoplastic process, which will continue automatically, even though the virus is eliminated. If, on the other hand, 'early' viral antigens are present in the tumour tissue and detected by analysis, then there is active viral synthesis. If, on the other hand, 'early' antigens are detected by serological methods, this means that there has been synthesis of viral protein of sufficient extent to stimulate antibody production; yet the virus may have been eliminated and viral synthesis may no longer be active. The demonstration, therefore, by Rawls and his colleagues (1972, 1973) of antibodies against early HSV_2 antigens in cervical carcinoma tissues, added to the presence of viral DNA in 30% of cases, lends powerful support to the supposition that HSV_2 is concerned with the etiology of this tumour. Many students, indeed, consider that the evidence is strong enough to be acceptable as a working hypothesis.

If the causative relationship between HSV_2 and cervical carcinoma were definitely established, the role of this virus in other cancers of the genital tract would appear more probable, especially in tumours of the prostate gland. Added plausibility would also be added to the tenuous link between HSV_1 and some cancers of the mouth and pharynx. Meanwhile, it is necessary to study the role of the human oncornaviruses in the causation of human neoplasia.

V Oncornaviruses and cancers of man

1 *General*

Should there exist in man a pattern of viral cancers analogous to those of chickens, mice and other animals, such as we have been considering in preceding chapters, we should expect to find: 1. a range of leukaemias and lymphomas, 2. a related series of hard tumours of the sarcoma range, caused by viruses, which might or might not be 'defective' requiring 'helpers', 3. of the latter a virus causing tumours of the bones such as osteo-sarcomas, and 4. a virus, which causes mammary carcinomas.

Man does indeed suffer from similar tumour systems, and, as established in earlier chapters, he does carry viruses related to those which cause these cancers in other animals, even non-human primates. It would not be unreasonable, therefore, to suppose that the human cancers were caused by these same viruses. However, if this is so, as with both the murine and bovine leukaemias, convincing proof of it has been exceedingly difficult to come by, though at times it has seemed near to being achieved. Epidemiological studies, as shown earlier in this chapter, do not indicate that cancer epidemics normally occur. We have, therefore, to seek our clues with the endogenous oncornaviruses, that are vertically transmitted and in normal circumstances harmlessly associated with the genetic apparatus of human cells. The two groups of oncornaviruses mainly involved are: 1. those that produce C particles and are responsible for causing leukaemias, lymphomas, sarcomas, osteo-sarcomas and sometimes carcinomas, and 2. the A and B particles, which are the cause of mammary carcinomas in mice. However, when in the latent or 'provirus' stage in the host genome, complete viral particles are not usually produced, though they are found quite often associated with neoplastic tissues. The oncornaviruses are a group on their own, which differ in important respects from other groups of viruses. We have seen with the herpesviruses, that they cause disseminated febrile diseases, such as are usually associated with viral infections; they then become latent in the body tissues, and only rarely cause cancers. The oncornaviruses have no such options; none is ever associated with the normal inflammatory type of disease, to which a name can be given; they are either latent or they cause cancer. Furthermore, when they replicate within a cell and produce complete viral particles, these are shed without cell destruction as occurs with all other viral groups.

2 *Transmission experiments*

Numerous attempts have been made to transmit human cancers including leukaemias to animals. Reviews, quoted by Gross (1970), have been published by Velich (1898), Brand (1902), Borrel (1907), Wolff (1907), Woglom (1913), Blumenthal (1962a, b, 1932), Ewing (1940) and Oberling (1954). These authors reported total lack of success. A number of workers also have reported apparent successes, which are critically reviewed by Gross (1970). Some of the results reported are impressive, but not scientifically convincing, and it must suffice here to reproduce the conclusions reached by Gross as a result of his review:

> The critical reader will notice that unquestionable and reproducible evidence has not yet been furnished. Additional and confirmatory data are needed. The induction of neoplasms, including leukaemia, with filtrates presumably containing human oncogenic viruses, would be far more convincing if mice of established inbred strains, bred under careful supervision, and known for their low incidence of spontaneous tumours, were employed for bio-assay, instead of using unpedigreed mice with unknown incidence of spontaneous tumours, purchased from commercial dealers. Adequate litter-mate controls, inoculated simultaneously with heated, or otherwise inactivated, human tumour extracts would be essential.

The situation, as stated by Gross in 1970, remains virtually unaltered today, but one apparently successful attempt to transmit human leukaemia to macaque monkeys and baboons deserves more complete mention.

In this study, reported by Lapin *et al.* (1975), the authors certainly induced a severe epidemic of leukaemia in their experimental animals following the injection of filtered material from human leukaemic patients. The experiments were properly controlled, and the disease induced was transmitted in the baboons naturally both horizontally and vertically. The authors, men of international repute, claimed that they had, indeed, transmitted human leukaemia and proved the involvement of a type C oncornavirus. Critics have asserted that this was not proved and they they had, in fact, activated a latent macaque or baboon leukaemia virus, which was not of human origin. This is difficult to accept. However, let us recount the story, as told by Professor Lapin and his colleagues.

Boris Lapin was at that time the director of the Institute of Experimental Pathology and Therapy, a well-known and respected research centre, delightfully situated on the north-east coast of the Black Sea, in Azerbaijan, USSR. At this centre for a great many years free-ranging colonies of baboons and other monkeys have been maintained for medical research. Experiments on the transmission of human leukaemia to stumptail monkeys (*Macaca arctoides*) and baboons (*Papio hamadryas*) were reported in a number of Russian journals from 1967 onwards, and in English by Lapin (1975) and Lapin *et al.* (1975). The monkeys regularly developed malignant lymphomas, following the injection of human leukaemic blood. The publication by Lapin *et al.* (1975) reports an extension of this work. The monkeys were injected with 20–30 ml of whole blood and the same amount of filtered plasma intra-peritoneally taken from 32 human patients suffering from different forms of leukaemia. Two hundred and fifty one macaques were used of both sexes mostly 2–4 years old. The experiments were controlled by injecting 6 monkeys with heat-inactivated human leukaemic blood and 8 monkeys with healthy human blood. After an incubation period of 10 days to 3 months (occasionally as long as 15 months), malignant lymphoma developed of severe and mild forms in different cases; the disease ran a wave-like course with relapses and remissions. The disease was aleukaemic, but both spleen and lymph nodes were greatly enlarged. There were usually 2–3 relapses before a fatal termination. When the blood had been taken from a human donor with lymphocytic leukaemia, the severe type of disease appeared in 40–46% of the experimental monkeys, but the incidence of the severe type of disease was much lower when the donors suffered from granulocytic leukaemia.

The disease was transmitted by passage from sick monkeys to healthy by means of whole blood, cell free material and spleen cells. In these passages, malignant lymphoma developed from both filtered and unfiltered material. The disease was transmitted from monkey to monkey for a total of 13 passages. In a total of 186 monkeys injected with material from acute forms of leukaemia, 39 developed the severe form of disease and 74 the mild; 43 remained healthy and 30 died shortly after receiving the injection. Of 43 monkeys injected with material from patients with chronic granulocytic leukaemia, 18 developed the severe form and 6 the mild. Of 22 monkeys injected from patients with chronic lymphocytic leukaemia, 7 developed the severe form and 11 the mild form. As with the primary injections of human blood, the passage experiments were controlled by injection of heat-inactivated blood from sick monkeys, non-infected monkey blood and non-infected monkey spleen cells. C type viral particles were readily demonstrated by electron microscopy in spleen, lymph nodes and bone marrow. Infected cells were also grown in tissue culture and produced viral particles.

Experiments with the baboons (*Papio hamadryas*) had also been previously reported in Russian literature. The results were even more interesting than with the macaques, because some of the infected baboons became naturally infected from contact. In this way a leukaemia epidemic was started in the normal baboon colony at the institute. Initially, 22 baboons were injected with human leukaemic blood. Results were similar to those with the macaques, and a further 21 baboons were used for 3 serial passages. C particles were similarly demonstrated in cells of the blood and bone marrow. The cases, which arose accidentally from contact transmission, appeared some 2 years after the sick and healthy baboons were housed together. At the time of their report, over 100 cases had arisen in this way and 49 baboons had died.

The virus, which caused the lymphomas in the stumptail monkeys and baboons, proved to be immunologically distinct from any C type oncornaviruses hitherto described, including simian sarcoma virus (SSV). Virus was also found in urinary sediments from the infected baboons, and was capable of transmitting the disease on injection into healthy animals. Two cases of congenital malignant lymphoma were also found in newborn baboons, showing that vertical transmission was occurring as well as horizontal.

A report of this nature from an institute of worldwide renown under the signature of a scientist of the standing and international repute of Professor Lapin cannot be lightly dismissed. Three possible explanations can be advanced: 1. that endogenous monkey and baboon viruses had been activated by the injections of human leukaemic material, 2. that a human leukaemic virus had hybridised with monkey and baboon leukaemic viruses, and 3. that the malignant lymphomas were caused by viruses, which caused the human leukaemias. It is difficult to believe that endogenous viruses had been provoked into activity in both groups of monkeys; that this might have happened in one of the two genera might be possible, but in both stretches the imagination too far. If so, why should the heightened virulence of the provoked virus be maintained during subsequent passages and under conditions of natural transmission? If this explanation is dismissed, as it surely must be, then either a human virus became hybridised with endogenous simian viruses, or else it caused the disease on its own. In either of these events, it is evident that a human leukaemia virus was active in the leukaemia patients, though not in healthy persons.

3 *Viral particles, DNA and antigens in human leukaemias and lymphomas*

Oncornaviruses in tumour cells are mostly concealed in the host DNA as 'virogenes' and only exceptionally produce intact viral particles to be found in the cell cytoplasm or free in the blood or tissues. The absence of such particles does not, therefore, teach anything, though their presence in tumour tissues or in the blood of cancer patients, while absent in healthy persons, is evidence for the association of the virus with the cancer. The presence of viral DNA or antigens in tumour tissue is better evidence than the presence of viral particles, provided that these substances are absent from healthy tissues or the amounts are much higher in the diseased ones.

It is undoubtedly the viral DNA in the host genome, which transforms cells either by stimulating the host gene responsible for its own DNA replication or switching off the host gene which regulates such replication. It is not surprising, therefore, that the search for C particles in tumour tissues has been somewhat fruitless, particularly where endogenous viruses are involved. Nevertheless, viral particles sometimes are found. The only really reliable evidence for the presence of C viral particles in human tumours is that presented by Dmochowski and his colleagues at the M. D. Anderson Hospital and Tumor Institute in Houston, Texas. Dmochowski *et al.* (1965) examined a large number of specimens from human leukaemia and lymphoma patients and found virus-like particles, similar to those which are found in leukaemic mice and rats. The particles were found in the cell cytoplasm budding in the characteristic manner of RNA viruses, in vacuoles in the cytoplasm or in intercellular spaces. Ultrathin sections of lymph nodes, obtained by biopsy from 16 patients with acute lymphatic leukaemia, were examined by the electron microscope, and typical virus-like particles were found in 8. Similar studies of patients with chronic lymphatic leukaemia revealed particles in only one specimen of 10, and 2 of 7 from patients with monocytic leukaemia. Studies of biopsy lymph nodes from 38 patients with lymphosarcoma or reticulum-cell sarcomas revealed particles in 11. Furthermore, when ultrathin specimens of centrifuged blood pellets from 16 patients with different forms of leukaemia were examined, 3 were found to contain C particles. These striking results provide strong evidence of the presence in these patients of human leukaemia and sarcoma viruses, and the similarity to results with chickens, mice and rats suggests a causal relationship between the viruses and the disease.

Leukaemic human cells have also been grown in tissue culture and examined for the presence of viral particles. Although suspect particles have been described, the results are too conflicting for importance to be attached to them.

The search for viral DNA sequences in human tumour and leukaemic cells is briefly described in the Virus Cancer Program (1975, pp. 19–20) report under the heading 'Evidence for Human RNA Tumor Viruses'. Little is added in subsequent reports (1976–1978). This review states that a variety of evidence suggests that human cells, like animal cells, contain RNA viral material, and that this is involved in the initiation of malignant growth. In the normal human cell, it is argued, the viral DNA sequences are repressed. 'If they have anything to do with cancer they should become activated when the cell undergoes transformation to the malignant state'. At this time, therefore, the cell should contain appreciable amounts of typical products of RNA virus genes, both RNA and viral protein molecules, including the viral enzyme RDDP. Such products have been sought, especially in breast cancer (see below) and acute myelogenous leukaemia (AML).

In the leukaemia studies, the evidence is convincing. Particles have been found in the leukaemic cells of AML patients, which contained both the characteristic viral RNA and the enzyme RDDP. The enzyme has been purified and studied extensively. It proves to be immunologically similar to RDDP from 2 primate viruses, gibbon ape leukaemia virus (GaLV) and the simian sarcoma virus (SSV). The RNA, too, proved to be closely related to that of these same viruses and more distantly to mouse leukaemia RNA; the closest relationship is to SSV. Further tests showed that human AML cells also contained viral antigens related to those of the SSV/GaLV group. The review continues to state that studies of SSV show that it is not a *bona fide* woolly monkey virus, and hints that the sole woolly monkey from which it was isolated might well have acquired it from a human source.

Succeeding volumes of the Virus Cancer Program do not greatly enlarge on this theme, though (1978, p. 26) it is stated that cell cultures from the human lymphoid cancer, known as Hodgkin's Disease, have been studied. These studies reveal immunological similarities between certain viral antigens and the viral enzyme reverse transcriptase with those of GaLV.

4 *Human mammary tumours*

Only in mice have mammary tumours been fully proved to be of viral origin. Four important points need to be recalled in respect of the mammary tumours of mice: 1. they are normally transmitted to the offspring by virus present in the mother's milk, 2. their development is dependent on endocrine factors, and this makes it difficult to grow the tumour cells in tissue culture, 3. the virus involved is an oncornavirus, but of a different group from the leukaemia and sarcoma viruses, the particles being known as A and B, and 4. infection, though normally transmitted horizontally to infant mice in the milk, can also be 'carried' by the male and transmitted venereally to the female. There is much inferential evidence that human mammary tumours are caused by a virus similar to the mouse mammary tumour virus (MMTV), although the human tumour differs histologically from that caused in mice by the Bittner Virus.

Particles resembling the B particles of MMTV have been demonstrated in the milk of normal healthy women by Feller and Chopra (1968), and by Dmochowski *et al.* (1969) and by the same authors in tumour tissues from breast cancers, but not in normal breast tissues or in non-malignant breast tumours. These findings suggest that B type oncornaviruses may be commonly carried by women in their milk, and that they may cause malignant changes, if there are glandular disturbances or other oncogenic influences present. A review of the situation with regard to viral DNA sequences in human breast cancer tissues (also in leukaemia, lymphoma and sarcoma tissues) is given by Bowen *et al.* (1974). They indicate common nucleotide sequences between mouse mammary tumour virus genome and DNA from human breast cancer tissue. Schlom *et al.* (1971, 1972) have further shown that particles present in human milk, similar to type B virus in mouse milk, contain reverse transcriptase and RNA characteristic of oncornavirus.

Bowen *et al.* (1974) have also provided strong evidence in their own work that cells from human breast tumours contain antigens related to those present in MMTV. They believe their studies show that 'human breast cancer is viral in origin and that the human breast cancer virus is antigenically related to the mouse mammary tumour virus'. The evidence, which is of a technical nature, is impressive and justifies the authors' conclusions; those interested should consult the review quoted. The Virus Cancer Program (1975, p. 20) report states the situation briefly, and confirms what has been written above. Normal human milk and cells from human breast tumours contain identifiable particles that contain typical oncorna B virus RNA and the viral enzyme RDDP. Benign breast tumour tissues have no such particles, RDDP or RNA homologous to MMTV or any other suspect viruses.

The evidence for a viral causation of human breast cancer, though inferential, is thus very strong, and the virus implicated in most cases is likely to be related to MMTV. However, the issue has been made more complicated by more recent studies. The Virus Cancer Program (1976) report stated that further and virtually conclusive evidence had been shown by workers in New York and Texas that malignant breast tumours contained viral antigens related to MMTV. However, strong evidence had become available that some malignant breast tumours contained antigens belonging to the Mason–Pfizer Monkey Virus (MPMV), a virus which, as will be recalled, had been isolated from a spontaneous mammary adenocarcinoma of a rhesus monkey; this virus has been assigned to the oncorna D virus group. Supporting evidence for the involvement of MPMV was reported in the Virus Cancer Program (1977) report, and suggested that in some breast tumours both MPMV and MMTV might be involved. In the Virus Cancer Program (1978) report, the finding of MPMV particles in artificially infected rhesus monkeys' milk of animals free from malignancy was reported.

At this stage, we must leave the intriguing problem of the involvement of oncornaviruses in human cancers, and turn our attention to the other viral groups commonly associated with animal neoplasias.

VI Papova–papilloma and adenoviruses in human cancers

It is evident that two groups of viruses, the herpes- and the oncorna-, are likely to be implicated in the more important human cancers. Besides these two groups, man is a known carrier of two other viral groups of known oncogenic potential: the papova–papilloma group, and 2. the adenovirus group. The human papilloma (wart) virus certainly causes benign papillomas and warts, and is more than likely to be concerned with malignant growths of the bladder. There is no convincing evidence that other members of the papova group or the adenoviruses are concerned with human malignancies. Here we may leave the matter, though it should be noted that polyoma, SV40-like viruses, and adenoviruses are known residents of the human body, and this may not be the end of the story.

VII Hepatitis B virus and cancer of the liver

An unexpected addition to the family of human cancers, believed to be associated with viral infection, came with the publication of a small booklet by Malpas and Melnick (1981) entitled 'Hepatitis B Virus and Primary Hepatocellular Carcinoma', after this text was prepared.

Primary carcinoma of the liver is an unusual neoplasm in most countries, but very common for unexplained reasons in certain parts of Africa and the Far East, though its distribution is only focal even in these areas. Its preponderance in endemic areas has been attributed to various causes, including the consumption of groundnuts or grains contaminated with aflatoxin. It was found, however that in high incidence areas there was also a high incidence of infectious hepatitis caused by hepatitis B virus, and a connection between the two came to be suspected. Extensive serological surveys showed that a high incidence of antibody to the virus occurred in these same areas, but not elsewhere. The coincidence is so striking that an etiological role of the virus in the cancer is not seriously doubted. It is intended to test this by immunising a proportion of the populations at risk, to discover whether protection against the virus will decrease the risk from hepatic carcinoma.

There are unfortunately certain difficulties, since it has not so far been possible to grow hepatitis B virus in culture, so that immunising antigen must be harvested from persons carrying infection. The virus is at present assigned to a miscellaneous group, but does not belong to the four viral groups, papova-, adeno-, herpes-, and oncorna-, hitherto associated with carcinogenesis. Its existence first became known, because it caused hepatitis in persons receiving injections of vaccines, such as yellow fever, in which human serum had been used as a diluent. Serum donors were, themselves, in good health, and the virus was obviously a 'passenger', as with other oncogenic viruses.

Evidence as to how it was transferred naturally from person to person was slow in coming. It is, however, not known that it can be transmitted either horizontally or vertically, vertical transmission probably occurring, as with herpesvirus, through the neonate.

VIII Summary and projection

It has been established beyond reasonable doubt that the major cancers which affect animals from mice to monkeys are directly caused by viruses, which belong to four or five of the numerous groups of these organisms. It has been further established that human neoplasias are intimately associated with at any rate two of these viral groups, herpesviruses and oncornaviruses. The inferential evidence that viruses of these groups actually cause the cancers, with which they are associated, is very strong and it is difficult to see how other than implied evidence can be obtained.

It is desirable to know more of the relationships of viruses with host cells, and why some few viruses stimulate abnormal activities in host cells, which would appear to have survival value neither for host nor virus. This must include an examination of viral genetics and antigen production, host responses and immunology, and the new influences which cause a harmless commensal virus to direct a host cell in an aberrant and harmful way. There would appear to be only one common link between those groups of viruses, which possess transforming ability, namely that they can all live in commensal association with their natural hosts, an association developed with some in evolutionary fashion over many millions of years. It would appear that under some conditions of stress, immunological embarrassment or trauma, the fine adjustment between host and virus becomes disturbed; the virus then induces the host cell to manufacture host DNA in excess, whereas in normal viral activity manufacture of viral DNA is stimulated. The result is uncontrolled host cell replication. This is unlikely to be the whole story, since it is known that host cell membranes are chemically altered by some viruses. These and other aspects of the problem require to be considered.

References

Balacesco. I. and Tovaru, S. (1936). Une observation authentique de transmission spontanée du cancer d'homme à homme. *Bull. du Cancer* **25**, 655–657

Beltrán, G., Baez, A. and Correa, P. (1966). Burkitt's Lymphoma in Colombia. *Am. J. Med.* **40**, 211–216

Blumenthal, F. (1926b). Beiträge zur Frage der Entstehung bösartiger Tumoren. Tumoren. *Deutsche med. Wochenschr.* **52**, 389–391; **52**, 435–438

Blumenthal, F. (1926b). Beiträge zur Frage der Entstenhug bösartiger Tumoren. *Deutsche med. Wochenschr.* **52**, 1283–1286

Blumenthal, F. (1932). Zur Frage der parasitären Krebsentstehung: Zum 100. Geburstage Ernst v. Leydens. *Zeitschr. f. Krebsforsch.* **36**, 30–144

Booth, C. C., Davies, J. L. P. *et al.* (1964). A case of African lymphoma demonstrated at the post-graduate medical school of London. *Brit. med. J.* **1**, 296–299

Borrel, A. (1907). Le problème du cancer. *Bull. Inst. Pasteur* **5**, 496–412; **5**, 641–662

Bowen, J. M., East, J. L., Fallen, F. T., Maruyama, K., Priori, E. S., Georgiades, J., Chan, J. C., Miller, M. F., Seman, G. and Dmochowski, L. (1974). Comparative morphology, immunology and biochemistry of viruses associated with neoplasia of animals and man. *In* (Kurstak, E. and Maramorosch, K., eds) 'Viruses, Evolution and Cancer'. New York and London: Academic Press

Brand, A. T. (1902). The etiology of cancer. *Brit. med. J.* **2**, 238–242

Burkitt, D. (1958). A sarcoma involving the jaws in African chidren. *Brit. J. Surg.* **46**, 218–223

Burkitt, D. (1962a). A lymphoma syndrome in African children. *Ann. roy. Coll. Surg. Eng.* **30**, 211–219

Burkitt, D. (1962b). A children's cancer dependent on climatic factors. *Nature* **194**, 232–234

Burkitt, D. (1962c). Determining the climatic limitation of a children's cancer common in Africa. *Brit. med. J.* **2**, 1019–1023

Burkitt, D. (1962d). A tumour syndrome affecting children in tropical Africa. *Postgrad. med. J.* **38**, 71–79

Burkitt, D. and Davies, J. N. P. (1961). Lymphoma syndrome in Uganda and tropical Africa. *Med. Press* **245**, 367–369

Centifanto, Y. M., Drylie, D. M., Deardourff, S. L. and Kaufman, H. F. (1972). Herpesvirus type 2 in the male genitourinary tract. *Science* **178**, 318–319

Chapman, D. S. and Jenkins, T. (1963). The Burkitt Lymphoma in Natal. A significant medical trail. *Med. Proc. (Johannesburg)* **9**, 320–331

Diehl, V., Henle, G., Henle, W. and Kohn, G. (1968). Demonstration of a herpes group virus in cultures of peripheral leucocytes from patients with infectious mononucleosis. *J. Virol.* **2**, 663–669

Dmochowski, L., Taylor, H. G., Grey, C. E., Dreyer, D. A., Sykes, J. A., Langford, P. L., Rogers, T., Shullenberger, C. C. and Howe, C. D. (1965). Viruses and Mycoplasma (PPLO) in human leukaemia. *Cancer* **18**, 1345–1368

Dmochowski, L., Seman, G. and Gallager, H. S. (1969). Viruses as possible etiological factors in human breast cancer. *Cancer* **24**, 1241

Epstein, M. A. (1970). Aspects of the EB Virus. *Adv. Cancer Res.* **13**, 383–411

Epstein, M. A. and Barr, Y. M. (1964). Cultivation in vitro of human lymphoblasts from Burkitt's malignant lymphoma. *Lancet* **1**, 252–253

Epstein, M. A. and Barr, Y. M. (1965). Preliminary observation on new lymphoblast strains from Burkitt tumours in a British and a Uganda patient. *J. nat. Cancer Inst.* **34**, 231–240

Epstein, M. A., Achong, B. G. and Pope, J. H. (1967). Virus in cultured lymphoblasts from a New Guinea Burkitt Lymphoma. *Brit. Med. J.* **2**, 290–291

Ewing, J. (1940). 'Neoplastic Diseases: A Treatise on Tumours'. Philadelphia and London: Saunders

Feller, W. F. and Chopra, H. C. (1968). A small virus-like particle observed in human breast cancer by means of electron microscopy. *J. nat. Cancer Inst.* **40**, 1250

Gerber, P. and Hoyer, B. H. (1971). Induction of cellular DNA synthesis in human leucocytes by Epstein Barr Virus. *Nature* **231**, 46–47

Gluckman, J. (1963). Multifocal lymphoma in South Africa. Its first observation in South Africa and in white children. *South African Cancer Bull.* **7**, 7–12

Goodheart, C. (1970). Herpesviruses and Cancer. *J. Am. med. Assoc.* **211**, 91–96

Grose, C. and Feorino, P. M. (1972). Epstein–Barr Virus and Guilain–Barré Syndrome. *Lancet* **2**, 1285–1287

Gross, L. (1970). 'Oncogenic Viruses'. 2nd edn, 991 pp. Oxford: Pergamon Press.

Henle, G. and Henle, W. (1966). Immunofluorescence in cells derived from Burkitt's Lymphoma. *J. Bact.* **91**, 1248–1256

Henle, G., Henle, W. and Diehl, V. (1968). Relation of Burkitt's tumour associated virus to infectious monocucleosis. *Proc. nat. Acad. Sci. USA* **59**, 94–101

Henle, W., Henle, G., Zajac, B., Pearson, G., Waubke, R. and Scriba, M. (1970a). Differential reactivity of human serums with early antigen induced by Epstein–Barr Virus. *Science* **169**, 188–190

Henle, W., Henle, G., Ho, H., Burtin, P., Cachin, Y., Clifford, P., de Schryver, A, de Thé, G., Diehl, V. and Klein, G. (1970b). Antibodies to Epstein–Barr Virus in naso-pharyngeal carcinoma, other head and neck neoplasms, and control groups. *J. nat. Cancer Inst.* **44**, 225–231

Kaplan, A. S. (ed.) (1973). 'The Herpesviruses'. New York and London: Academic Press

Katz, S. (1930). 'Henri Vadon. Vadon et le problème de la transmission du cancer'. Thèse, p. 106. Paris: Presses Universitaires de France

Khan, A. G. (1964). The multifocal lymphoma syndrome in African children in Kenya. *J. Laryng. (London)* **78**, 480–498

Klein, G. (1971). Immunological aspects of Burkitt's Lymphoma. *Adv. Immunol.* **14**, 187–250

Klein, G. (1972). Herpesviruses and oncogenesis. *Proc. nat. Acad. Sci. USA* **69**, 1056–1064

Klein, G., Clifford, P., Klein, E. and Stjernswärd, J. (1966). Search for tumor specific immune reactions in Burkitt Lymphoma patients by the immunofluorescence reaction. *Proc. nat. Acad. Sci. USA* **55**, 1628–1635

Klein, G., Klein, E. and Clifford, P. (1967). Search for host defences in Burkitt Lymphoma: membrane immunofluorescence tests on biopsies and tissue culture lines. *Cancer Res.* **27**, 2510–2520

Lapin, B. A. (1975). (*1*) Possible ways viral leukaemia spreads among the hamadryas baboons of the Sukhumi monkey colony. *and* (2) The embryologic and genetic aspects of an outbreak of leukaemia among hamadryas baboons of the Sukhumi monkey colony. *In* (Ito, Y. and Dutcher, R. M., eds) 'Comparative Leukaemia Research 1973: Leukaemogenesis'. University of Tokyo Press/Basel: Karger

Lapin, B. A., Yakovleva, L. A., Indzhiia, L. V., Agrba, V. Z., Tsiripova G. S., Voevodin, A. F., Ivanov, M. T. and Djatchenko, A. G. (1975). Transmission of human lukaemia to non-human primates. *Proc. roy. Soc. Med.* **68**, 141–145

Lecène, P. and Lacassagne, A. (1926). Une observation d'inoculation accidentale d'une tumeur maligne chez l'homme. *Ann. anat. path.* **3**, 97–112

Levine, P. H., Ablashi, D. V., Berard, C. W., Carbone, P. P., Waggoner, D. E. and Malan, L. (1971). Elevated antibody titers to Epstein–Barr Virus in Hodgkin's Disease. *Cancer* **27**, 416–421

Levy, S. B. (1974). Cat leukaemia: a threat to man? *New England J. Med.* **290**, 513–514

Malpas, P. H. and Melnick, J. L. (1981). 'Hepatitis B Virus and Primary Hepatocellular Carcinoma'. Basel: Karger

Miller, G., Niederman, J. C. and Andrew, J. L. (1973). Prolonged oropharyngeal excretion of Epstein–Barr Virus after infectious mononucleosis. *New England J. Med.* **288**, 229–232

Nahmias, A. J. (1974). The evolution (evovirology) of herpesviruses. *In* (Kurstak, E. and Maramorosch, K., eds) 'Viruses, Evolution and Cancer'. London: Academic Press

Nahmias, A. J. and Dowdle, W. R. (1968). Antigenic and biologic differences in herpesvirus hominis. *Proc. med. Virol.* **10**, 110–159

Oberling, C. (1954). 'Le Cancer'. 7th edn. Paris: Galliard

O'Conor, G. T. (1961). Malignant tumours in African children. II A pathological entity. *Cancer* **14**, 270–283

O'Conor, G. T. and Davies, J. L. P. (1960). Malignant tumours of African children with special reference to malignant lymphomas. *J. Paediat.* **56**, 526–535

O'Conor, G. T., Rappaport, H. and Smith, E. B. (1965). Childhood lymphoma resembling Burkitt tumour in the United States. *Cancer* **18**, 411–417

O'Conor, G. T., Rappaport, H. and Smith, E. B. (1965). Childhood lymphoma resembling Burkitt tumour in the United States. *Cancer* **18**, 411–417

Pagano, S. (1974). The Epstein–Barr viral genome and its interactions with human lymphoblastoid cells and chromosomes. *In* (Kurstak, E. and Maramorosch, K., eds) 'Viruses, Evolution and Cancer'. New York and London: Academic Press

Pulvertaft, R. J. V. (1964). Cytology of Burkitt's tumour (African lymphoma). *Lancet* **1**, 238—240

Rapp, F. and Koment, R. W. (1974). Herpesvirus etiology of abnormal growth. *In* (Kurstak, E. and Maramorosch, K., eds) 'Viruses, Evolution and Cancer'. New York and London: Academic Press

Rawls, W. E., Adams, E. and Melnick, J. L. (1972). Geographical variation in the association of antibodies in herpesvirus type 2 and carcinoma of the cervix. *In* (Biggs, P. M., de Thé and Payne, L. N., eds) 'Oncogenesis and Herpesviruses'. Int. Agency Res. Cancer, Lyon

Rawls, W. E., Adams, E. and Melnick, J. L. (1973). An analysis of seroepidemiological studies of Herpesvirus type 2 and carcinoma of the cervix. *Cancer Res.* **33**, 1477–1482

Scanlon, E. F., Hawkins, R. A., Fox, W. W. and Smith, W. S. (1965). Fatal transplanted Melanoma: a case report. *Cancer* **18**, 782–789

Schlom, J., Spiegelman, S. and Moore, D. H. (1971). RNA-dependent DNA polymerase activity in virus-like particles isolated from human milk. *Nature* **23**, 97

Schlom, J., Spiegelman, S. and Moore, D. N. (1972). Detection of high molecular weight RNA in particles from human milk. *Science* **175**, 542

Simons, P. J. and Ross, M. G. R. (1965). The isolation of herpesvirus from Burkitt tumours. *Europ. J. Cancer.* **1**, 135–136

Southam, C. M. and Moore, A. E. (1958). Induced immunity to cell homografts in man. *Ann. NY Acad. Sci.* **73**, 635–653

Southam, C. M., Moore, A. E. and Rhoads, C. P. (1957). Homotransplantation of human cell lines. *Science* **125**, 158–160

Ten Seldam, R. E. J., Cooke, R. and Atkinson, L. (1966). Childhood lymphoma in the territories of Papua and New Guinea. *Cancer* **19**, 437–446

de Thé, G. (1972). Virology and immunology of naso-pharyngeal carcinoma: present situation and outlook: A review. *In* (Biggs, P. M., de Thé, G. and Payne, L. N., eds) 'Oncogenesis and Herpesviruses'. Int. Agency Res. Cancer, Lyon

Velich, A. (1898). Beitrag zur Frage nach der Übertragbarkeit des Sarcomes. *Wiener med. Blätter* **21**, 711–712; 729–731

Virus Cancer Program (1975–1978). Washington: US Dept. of Health, Educational Welfare, Bethesda, Md.

Weber, F. P., Schwartz, E. and Hellenschmied, R. (1930). Spontaneous inoculation of melanotic sarcoma from mother to foetus. *Brit. med. J.* **1**, 537–539

Woglom, W. H. (1913). 'The Study of Experimental Cancer. A Review'. New York: Columbia University Press

Wolff, J. (1907). 'Die Lehre von der Krebskrankheit von den ältesten Zeiten bis zur Gegenwart'. Jena: Fischer

Woodall, J. P., Williams, M. C., Simpson, D. I. H. and Haddow, A. J. (1965). The isolation in mice of strains of herpesvirus from Burkitt tumours. *Europ. J. Cancer* **1**, 137–140

9
Cancer and Immunity

I General

In a study of the infectious causes of cancer, it is logical at some stage to consider whether the host's immune processes are involved and, if so, whether they play any significant part in the disease process, or could be of value in diagnosis, prophylaxis or therapy. A certain amount has appeared in previous chapters about the antigens of certain specific viruses in relation to the methods used to detect their presence, either as complete virions or as viral genes present in the genome of the host. It is now necessary to survey the immune problem in cancer more systematically.

This aspect of the cancer problem is admitted to be of importance both by the somatic and viral theorists. The somatic theory, as has been seen, attributes neoplastic growth to errors of DNA copying in somatic cells and failure of the natural repair processes. Such errors are occurring naturally and continually, and the faulty cells either fail to survive or are recognised as alien and destroyed by the host's normal immune mechanisms. There is more likelihood of such cells surviving in circumstances in which: 1. the host suffers from a predisposing genetic defect, 2. the immune system is embarrassed because of some other cause, which could be infection, trauma or even psychological stress, and 3. the weakening of the immune system as age advances. The neoplastic cells, besides their ability to proliferate at the expense of the parent tissue, possess certain properties that are alien to it; they synthesise as part of their aberrant behaviour potentially antigenic substances, such as foeto-proteins which are characteristic of cells at an earlier stage of development. There are undoubtedly antigens, which distinguish cancer cells from normal somatic cells, and which are distinct from those developed against infecting viruses, if such exist. These antigens are properties of cell membranes, proteins and glycoproteins, and any immunity developed is, therefore, against the rogue cell and not against any infecting organism. Once cancer immunities come to be discussed, the role of cell membranes in the development of the disease comes to be important; hitherto, the role of nuclear structures has been studied virtually to their exclusion.

Membrane changes are recognised to be of importance in virus induced cancers, as well as in those that are not supposed to be associated with viral infections. RNA viruses become attached to cell membranes, where they proliferate by budding, and cause changes in the chemical composition and properties of the membranes. This and other aspects of the immunology of neoplasia, where a virus is involved, are discussed by George Klein (1974) in chapter 17 of Kurstak and Maramorosch entitled 'Immunological Patterns of Virus-transformed Neoplastic Cells'. Cairns (1978) discusses the 'Immune Response to Cancer Cells' with his usual lucidity and brevity in only two pages. He finds that the immune system functions as a protective device against cancer caused by viruses, but cannot retard the process of chemical carcinogenesis. A most informative work on cancer immunity is *Man, Cancer and Immunity* by Alistair J. Cochran (1978). This work is less well known than it deserves, probably because the author has found difficulty in presenting a logical sequence of arguments, leading to useful speculations and deductions. It is, however, fully documented and provides the basic information for an understanding of the subject. The introduction by Klein (1978) is worthy of study.

II The nature of the neoplastic lesion

Cochran makes an interesting distinction between the pathology of *in situ* tumours and metastasising tumours. He finds the former to contain an inflammatory element in that the tumours contain numerous immunologically competent cells associated with them, whereas such cells are absent from secondary tumours. This, of course, is in conformity with the better results of treatments, in cases where a tumour has not begun to spread to new sites. It also contradicts earlier conceptions of neoplasms as being non-inflammatory, indicating furthermore that primary neoplastic tissues are recognised as alien and provoke a host reaction against them. Cochran (1978, p. 85) describes the situation as follows:'*In situ* malignancy is commonly associated with a dense infiltrate of inflammatory cells while invasive tumours often have a much reduced reactive component confined to the tumour periphery rather than infiltrating it'. The malignant tumour may, then, by definition be a non-inflammatory lesion, but its presence may provoke an inflammatory reaction. This being the case, the tumour must contain antigens capable of provoking this reaction.

That there should be a response at all is of importance to our enquiry, and the nature of the response could have a bearing on the question as to whether there was a viral component involved, since the antigens could be derived either from the deranged tumour cells, or from some viral components or from a combination of the two. The distinction between the exogenous and the endogenous virus is in this respect of importance, since the endogenous virus will have a host tolerance, acquired during foetal life, whereas the exogenous virus will undoubtedly provoke an immune reaction. Similarly, if neoplastic cells are indeed derived from somatic cells in which DNA copying has been defective, there does not appear to be any good reason why they should provoke an immune response, in spite of their irrational behaviour. In rare human cases, the immune challenge leads to spontaneous resolution of the tumour; perhaps, this is more common than supposed, and many small tumours, not clinically recognised as such, are automatically eliminated. Such resolution, as we have seen, commonly occurs with some virally induced animal tumours, such as the papillomas of cottontail rabbits and cattle.

Where viruses are involved, an explanation for this phenomenon is easily found. Resident viruses, such as *H. simplex* or varicella do commonly become pathogenic, if the host is exposed to immunological stress; or they may become active if a second 'helper' virus is acquired. If viruses are not involved, the reaction must be in the nature of an 'auto-immune' response; alternatively, the tolerogenic memory of certain foetal proteins, known to be associated with cancers, must have weakened. This response is developed against cell surface or membrane antigens, amongst which are included a number of so-called 'oncofoetal antigens'. Cochran lists a number of such, which are clearly derived from the tumour cells but which are absent from normal cells.

We have seen in earlier chapters how, by a variety of elegant and sophisticated techniques, the involvement of viruses with malignant tumours can often be demonstrated, though their presence does not necessarily prove an etiological role in the neoplastic process. For example, antibodies to EBV are present in 90% of healthy persons, but are present in all sufferers from Burkitt's Lymphoma and in much higher titre. It is evident that immunological restraints prevent the development of the tumour, except in those geographical areas where it is prevalent and some other contributory factor must be present. In cervical carcinoma of women, evidence of the presence of *H. simplex* 2 is present in some 60% of sufferers, but rare in women who are not affected; here again the evidence for a viral etiology is good but not conclusive. One would have thought, that it would not be difficult to prepare a vaccine and compare results in comparable groups of women at equal risk. In human breast cancer, there is present a number of antigens absent from non-neoplastic breast tissues; these include one that cross-reacts with mouse mammary tumour virus (MMTV), and indeed both milk and serum from women with breast cancer can neutralise MMTV. Serological relationships have also been established between antigens present in human leukaemias and those of RNA tumour viruses in animals, *vide* Metzgar *et al.* (1976).

III Stimulation of the immune response

Available evidence, thus, clearly shows that at some stage neoplastic growths are recognised as alien by the immune defences of the host, which with some tumours and under certain conditions, that are not known, may succeed in eliminating them. Something is known of the antigens involved, some of which appear to suggest that viruses may be present in tumour tissues while absent from those that are healthy; these will be studied further in a later section. Meanwhile, it may be reasonably asked, whether measures taken to stimulate the immune system would not assist cancer therapy? Such might be supposed likely to be helpful, especially in primary tumours where immune defences are already active. The short answer is, that in some cases results are effective, but are always unpredictable and in other comparable cases may be unhelpful or deleterious. Immunological adjuvants could either stimulate the defence mechanisms or absorb those that were already present with opposite results. The problem is discussed by Cochran (1978, p. 167 *et seq.*). A number of different means of stimulating the immunological system have been tried and some spectacular successes have been achieved, particularly in children suffering from leukaemia. The most commonly used agent is the attenuated tubercle bacillus BCG (Bacillus Calmette-Guérin), commonly used to immunise young persons against tuberculosis. Other agents have also been used and successes have been achieved with some other cancers, such as melanoma, which usually proves rapidly fatal. The immune processes against these cancers are both cell-mediated and humoral, and are consistent with the view that they are directed against the 'rogue' cells of the cancer and not against any infectious agent.

In animal cancers, there is little difficulty in inducing either active or passive immunity with vaccines or immune sera, when the causal organism is known. For Marek's Disease in chickens, a reliable vaccine is in commercial production; apart from this, a number of vaccines have been prepared against various viruses, which successfully prevent the development of the disease under experimental conditions. There seems little doubt that, were viral causes to be determined for the commoner cancers of the human race, the viruses could without difficulty be manipulated so as to give protection against some of the commoner and more distressing cancers, such as the leukaemias and cancers of the breast. It is for these reasons so desperately important that the somatic *versus* virus theories of cancer should be resolved. If the viral theory is correct, some at any rate of these diseases could be eliminated; if the somatic theory is correct, there is little likelihood that the incidence of these diseases can be diminished, though treatment routines may be improved.

IV Antigens of neoplastic diseases

Viral antigens play an important part in the prevention of diseases caused by viruses, and thus in preventing the development of any cancers they induce. However, as stated above, the host is unlikely to develop a useful immune response against endogenous viruses, as with the murine and probably the human oncogenic viruses, only against exogenous viruses such as the avian and feline. However, the presence of an endogenous oncogenic virus does not mean that the host will develop cancer; indeed, as with the adenoviruses in man and some herpesviruses in monkeys, they may only be capable of causing cancer in alternative animal species. If an exogenous oncogenic virus should be acquired, cancer will develop, so that the effect of the vaccine is to prevent infection with the virus not to prevent the development of neoplasia. Control of tumour growth and development, on the other hand, evidently lies predominantly with the immune response to the neoplastic cell's own 'membrane antigens', which are not viral antigens; however, the nature of cell membrane antigens is profoundly influenced by viral infections. The nature of the viral influence on host cell membranes is discussed at length by Myron Essex (1974). She writes:

> The immune response to oncornavirus-induced cell membrane antigens is probably the primary immune mechanism for controlling tumor growth. The immune response to this group of antigens is distinct from, and does not necessarily follow, the responses to either virus envelope or core antigens. The immune response directed to the cell membrane antigens can probably both block tumor outgrowth and induce regression of some established tumors. In the case of spontaneous and induced fibrosarcoma and leukemia of outbred cats, the humoral antibody response to the oncornavirus-induced cell membrane antigen is closely correlated, in the inverse sense, with tumor growth. Cats that develop high antibody titers either develop no visible tumors or develop tumors that regress, while cats that fail to develop high antibody titers die with progressing fibrosarcoma or leukemia. Both cell-mediated and humoral responses appear to contribute in a favourable sense to this activity.

The important point is here made that anti-tumour immunities are developed against cell membrane antigens, the development of which may be virus induced but the antigens are not viral antigens. This means that there is in the neoplastic cell membrane a new product, which is native neither to the parent cell nor to the virus, and this can explain how both host cell products and tolerogenic viruses come to provoke immune responses, as if they were alien which in combination in an estranged form they are.

George Klein (1974) comments that the presence of proliferating oncornavirus within the host cell does not necessarily lead to its destruction. It is not, therefore, incompatible with cell proliferation, normal or neoplastic. With all known DNA viruses, the situation is different in that the presence of proliferating papova-, adeno-, or herpesviruses damages the host cells beyond recovery and is, therefore, incompatible with proliferation. In this case, it must be supposed, in the case for example of EBV-transformed lymphoblastoid cell lines, either that viral antigens are not made ('non-producer lines'), or, if they do, they make them by throwing off abortive 'side-lines' that enter the productive cycle and are thereby eliminated automatically from the mainstream of cell proliferation. The author suggests that there are two lines of DNA-virus induced antigens, one of which is not compatible with cell proliferation. There is a second large category of 'viral or virally determined products', which is regularly associated with virally transformed cells and is compatible with continued cell multiplication; these 'products' include intranuclear and membrane antigen. In this context, the pheonmenon noted in an earlier chapter, that B lymphocytes can only be grown in tissue culture, if they carry EBV DNA, is of interest.

Two outstanding points should be noted with regard to the intra-nuclear antigens: 1. that some are products of host gene activity, not of viral genes, and 2. that some such may be responsible for initiating the neoplastic process. Klein tends to the view that it may well be cell membrane changes, which induce the neoplastic changes, rather than direct gene involvement. Writing in relation to polyoma in mice, he states the case as follows:

> The concept that polyoma-induced TATA (tumour-associated transplantation antigen) reflects an essential membrane change for neoplastic behaviour is attractive for several reasons. It is widely believed that membrane receptors are profoundly involved in growth control. Cell contact-dependent signals involved in the repression and derepression of DNA synthesis are probably triggered by membrane receptors. Membrane receptors are also involved in transmitting most hormone signals that stimulate or restrict cell proliferation. A membrane change that is regularly associated with a certain kind of virally induced neoplastic transformation is therefore a reasonable candidate for the role of *the* key change that has put the normal receptor system out of function. Damage to the relevant receptor may unlatch a free-wheeling cycle of DNA synthesis.

The author continues by showing that inoculation of young mice with virus-induced TATA leads to effective resistance against the development of polyoma on challenge. He reiterates furthermore the widespread belief that *all* virus induced tumours have new membrane antigens. He does not, however, commit himself to the belief that all neoplastic tumours are virus-induced, since he appears to believe that similar membrane changes could be induced by carcinogenic chemicals. I must again quote from this highly articulate author in his Foreword to Cochran's *Man, Cancer and Immunity* (1978). Klein writes:

> Yet there is increasing evidence that not all tumours are recognised by the immune responses of the host. And why should they be? Virus-induced tumours bear the same surface-associated antigens, as long as they are produced by the same virus. In this special situation surveillance has a clear target to focus on and, in the cases where the species had previous extensive contact with the virus, is aided by an immense pre-history of natural selection.
>
> Spontaneous tumours seem likely to represent a very different situation. These are tumours which arise without experimental interference and emerge at the end of a long progression which is now understood to be the gradual evolution of cellular independence from a variety of local and general restrictive influences including hormonal and, no doubt, immunological factors. If tumour-associated antigens do arise in slow-developing spontaneous tumours it would be expected that they would meet a strong selective pressure resulting in *low imunogenic or non-immunogenic tumours* [my italics] in contrast to the rapidly developing tumours induced by strong chemical carcinogens or powerful oncogenic viruses.

Klein's arguments are impressive. We started with two theories of cancer causation, the somatic and the viral. He produces a third, which some may think the most plausible in that it explains many anomalies and to some extent bridges the gap between the somatic and viral; or rather he demolishes both by claiming that cell transformation results neither from errors of DNA copying nor from the presence of viral genes in the host genome, but that the aberrant behaviour of cell DNA is caused by damage to or changes of the membrane receptors on the cell surface, however caused.

V The role of the cell surface in immunity

The most important contribution to the role of the cell surface in neoplastic processes is that of Rose Sheinin (1974), also to be found in the invaluable book *Viruses, Evolution and Cancer* edited by Kurstak and Maramorosch. Her arguments are supported by no less than 175 references to original literature, and her review is written in such lucid language that there is little difficulty in following the theme. Her analyses of the evidence, furthermore, are an exemplary model of what scientific deduction should be. I make no apology for quoting extensively from this able paper.

In her opening pages, Rose Sheinin lists the major functions mediated by the cell surface, quoting 9 major heads with 29 sub-heads. Of the latter, there is good evidence that some 26 are modified by neoplastic transformation of the cell. Amongst the modified functions are: 1. various enzyme activities, 2. transport systems 3. interaction with viruses, 4. interaction with agglutinins and antigenicity, including blood group substances, tissue specific antigens, foetal antigens, and immunoglobulins, 5. interaction with the immune system, including reactions with immune lymphocytes, macrophages, cytotoxic sera and complement, 6. growth contact inhibition, involving movement, mitosis, and progression through the cell cycle, 7. morphogenesis, cell recognition and adhesion, and 8. interaction with regulatory molecules, such as hormones. She postulates that specific molecules at the surface of cells perform all these functions, in some instances the postulate resting on only a few observations, but in others on data converted from hypothesis to established fact.

The working model of the cell surface is that proposed by Singer and Nicholson (1972). The entire 'plasmalemma' comprises a fluid lipid phase into which are inserted structural and functional proteins and glycoproteins. Little is known, however, of the ways in which the structure is ordered, although there is considerable information on the means of synthesis by the cell of neutral lipids, phospholipids, sterols, glycolipids, proteins and glycoproteins, substances which must be integrated to form a functioning plasmalemma. According to Singer and Nicholson (1972), the peripheral components of the cell membrane are directly or indirectly modified as a result of virus infection arising from: 1. entry of the virus into the cell, 2. exit of newly replicated virus from the cell, and 3. interference with the 'biogenesis' of the cell membrane. It is the latter, which is likely to be involved with oncogenesis.

Enveloped viruses, which include those that are oncogenic, do indeed have a profound effect on the synthesis and turnover of the plasmalemma. These viruses contain a core of nucleoprotein surrounded by the capsid (envelope) proteins, and are encased in a membranous envelope. The viral membranes, however, closely resemble those of the host cells, and the proteins and glycoproteins are encoded directly by the viral genome. Hence, in Sheinin's words: 'The surface biology and biochemistry of host cells are profoundly altered, when viral envelope proteins are inserted into the plasma membrane. This is reflected in an altered immuno-pathology'. Sheinin repeats the point made by Klein (above) that most instances of productive viral infection are irrelevant, because the affected cell dies. However, there are at least three types of virus-cell interaction in which host cells would survive, but biochemical alterations of the cell surface could be caused. These are:

1. Cells which have developed a carrier state and continue to bud off complete or defective viruses from the cell surface [e.g. oncornaviruses]; 2.

Abortively infected cells in which are expressed those factors determining synthesis and movement to the surface of viral envelope proteins, and 3. Cells infected with virus that is a mutant in some terminal function of the replication cycle, but which does express those functions that modify the cell surface.

Oncogenic viruses can, thus, function as donors of genetic information, which can modify cell surface functions, and it is known that such cells carry virus DNA as an integral part of their chromosomal DNA. Sheinin continues:

> We now know that virus transformation is a two-step process. Genetic transformation, which results from incorporation of virus genetic material into the chromosomal DNA, is necessary but not sufficient to produce a neoplastically transformed cell. It requires in addition phenotypic expression of a virus function which ultimately modifies the surface of the affected cell. It is the cell surface that embodies a key lesion of neoplastic transformation, for it mediates those phenomena of cell-cell interaction that underlie both normal and cancerous growth and development.

Rose Sheinin is clear that the neoplastic process is that, which is outlined above, and she quotes ample authority for her views. She does not, however, exclude the possibility that other agents apart from viruses could have a similar effect. While it seems evident that the virus genome is the repository of information for neoplastic transformation, it is not established whether specific molecules are continually elaborated and incorporated into the plasma membrane or whether the regulation of the normal metabolic pathways become chronically disturbed. If the 'hit and run' mechanism, discussed in earlier chapters, should be correct, then the latter view must prevail and a single exposure to a carcinogenic influence could start the oncogenic process; such could arise from drugs or radiation. If the former alternative should be correct, it is difficult to see how the oncogenic process could be maintained in the absence of active viral genes. Those interested would find a study of Rose Sheinin's review article rewarding.

VI The immune system and treatment routines

Cancers are treated by drug or ray therapy, or by a combination of the two. One might suppose, because of the special properties of cancer cells, that it should not prove too difficult to find chemical agents which would prove lethal to them, while sparing the healthy cells of the body. Much has indeed been achieved, and a great many cancer sufferers have had their lives prolonged by judicious courses of treatment. Such courses may, or may not, be combined with surgical removal of the tumours together with the associated lymphatic glands. Selection of the routine to be followed in each individual case is a matter for the specialists. Successes are more easily achieved with early primary cancers, in which there is a significant inflammatory reaction. Where there are secondary tumours, the outlook is poor and complete elimination of the neoplastic process is unlikely to be achieved. The best that can be hoped is prolongation of life usually with continuous drug therapy often accompanied by unpleasant side effects. It would appear that successes with early primary cancers depend to a great extent on the body's own immune mechanisms and, as suggested above, suitable measures to enhance this activity may be helpful.

VII Conclusion

This brief survey of the involvement of immune mechanisms in cancer leads to some important conclusions. The first conclusion is that they are involved and are important. Those, who believe that cancers all arise from infectious causes, may not be surprised, but they may be surprised to learn that the immunity is not developed primarily against the viruses they suppose to be the cause of the cancers. Nor is the immune reaction developed against the cancer cell as such. It is developed against new and alien products, proteins or glycoproteins, which replace those that are native in the cell membranes on the cell surface. They are new and alien and are thus antigenic; they are substances, neither somatic nor endogenous viral, to which the host has developed tolerance during foetal life. So, in primary cancers, the body defences can and do react against them. Though the antigens are new products, in cancers caused by viruses or in which viruses are involved they contain a viral element which can be traced and identified. Such elements can be traced and identified in some cancers, such as human leukaemias and breast cancer, which are not hitherto regarded as of viral origin; they give, however, a strong indication that viruses may be involved in such cancers. Immunological studies, furthermore, redirect attention from what is happening in the cell nucleus and its DNA to the cell surface, which may well prove to be of greater importance.

There seems to be little hope that immunisation procedures could help cancer sufferers, but *if* viruses are involved, protection by vaccines could be afforded against *exogenous* oncogenic viruses. Or it could be afforded against exogenous viruses, which act as 'helpers' or hybridise with endogenous viruses. The immune element in cancer is deserving of more attention than has so far been given to it. Cochran's book on the subject is worthy of study.

References

Cairns, J. (1978). 'Cancer, Science and Society'. San Francisco: Freeman
Cochran, A. J. (1978). 'Man, Cancer and Immunity'. London: Academic Press
Essex, M. (1974). The Immune Response to Oncornavirus Infections. *In* (Kurstak, E. and Maramorosch, K., eds) 'Viruses, Evolution and Cancer'. New York and London: Academic Press
Klein, G. (1974). Immunological Patterns of Virus-transformed Cells. *In* (Kurstak, E. and Maramorosch, K., eds) 'Viruses, Evolution and Cancer'. New York and London: Academic Press
Klein, G. (1978). Foreword to 'Man, Cancer and Immunity', by A. J. Cochran. London: Academic Press
Metzgar, R. S., Mohanakumar, T., Schäffer, W., and Bolognesi, D. P. (1976). Relationships between Membrane Antigens of Human Leukemic Cells and Oncogenic RNA Virus Structural Components. *J. exp. Med.* **143**, 47–63
Sheinin, R. (1974). The Cell Surface, Virus Modification, and Virus Transformation: *In* (Kurstak, E. and Maramorosch, K., eds) 'Viruses, Evolution, and Cancer'. New York and London: Academic Press
Singer, S. J. and Nicholson, G. L. (1972). The fluid mosaic model of the structure of cell membranes. *Science* **175**, 720–731

10
Epilogue

The first part of this book covers a period of more than seventy years, since Ellerman and Bang in 1908 and Peyton Rous in 1911 demonstrated that certain neoplastic diseases of chickens were transmissible by cell-free material, and, therefore, caused by viral infections. Seventy years, more than two human generations, is a long time considering the intensity of the research effort and the enormous sums of money spent on human and animal cancers. Yet we still cannot answer the question as to whether all cancers are the result of the presence of viral DNA in the host genome. We cannot, with complete confidence, assert this in relation to animal cancers, though the evidence strongly suggests that this is so. An uneasy feeling must remain that there could be other influences, which may pervert cell DNA besides viruses. In this case, while viral DNA is demonstrably the most important influence in animal cancers, other influences might be of greater importance in man because of the unnatural conditions of life which he has created for himself.

It is, however, admissible for scientists to propose hypotheses, based on adequate facts and subject to scrutiny and experimental tests; this approach was justified by Claude Bernard as long ago as 1865 in his classical *Introduction to the Study of Experimental Medicine,* in which the second chapter is entitled 'The A Priori Idea and Doubt in Experimental Reasoning'. The question must be considered, therefore, as to whether the evidence derived from the studies of animal cancers, together with those of some human cancers, is sufficiently strong to justify the hypothesis that all malignant diseases are ultimately the result of viral genes in host DNA?

The US Virus Cancer Program was inaugurated on the hypothesis that all cancers were the result of viral infections. The program was started in 1964 with a budget of $4 926 000 rising to some $60 million annually from 1975 to 1978. By 1975, it had ruefully to be admitted that the hypothesis must be modified; the search was no longer for a viral cause of cancer, but for the presence of viral genetic material present in and actively influencing the replicatory activities of neoplastic cells. It is not our object in scientific research to 'prove' our hypotheses, but to 'test' them; we try to prove ourselves wrong. All too often, a scientist will convince himself that the evidence has been so critically analysed that his hypothesis can have no flaws, only to find that, when put to the test, his hypothesis is invalid. This discovery often leads to a new line of investigation, whereby the problem will be solved, or to some new discovery of far-reaching importance. The experienced chemist, who purifies his product with meticulous care, finds his end product inert, while his inexperienced colleague achieves positive results; the active principle lies in the 'gunk', the contaminating chemicals, which in his inexperience he has failed to remove. It was amongst the 'gunk' of virological research, that the active principle of viral activity in relation to cancer was discovered, namely not the virus but the viral genes.

The original demonstration of the viral origins of some chicken cancers resulted from normal investigative procedures. The work of Ludwik Gross on the mouse cancers and succeeding investigators on the cancers of other animals have been mostly based on the hypothetical view that oncogenic viruses were involved, and confirmation has come from the eventual isolation of complete viral particles, enabling their study, classification and demonstration that they could reproduce neoplastic conditions in experimental animals. To the independent observer studying the reports of the Virus Cancer Program, it is evident that the modification of the original hypothesis from a search for a viral cause of cancer to that for viral genetic information caused a change of emphasis, which impeded the course of the investigations. After 1975, the main effort was devoted to the development of techniques for detecting the presence of viral genes in host DNA, and progress was impeded to such an extent that the program appeared to have failed in its objectives by the time that financial stringency reduced the support given to it.

While there is room for disagreement, there are evidently sufficient grounds to justify the proposal that all cancers, animal or human, are associated with the presence of viral DNA in the host genome. The reorientation of effort by the Virus Cancer Program necessitated not only the demonstration that viral DNA was present in the host genome, but also that the neoplastic condition was caused by its activity. This was the more difficult, because it was found that inactive viral DNA was present in virtually all body cells, including those of reproduction by which it was transmitted from generation to generation as if it were host DNA. This made more plausible the belief that co-carcinogens caused cancers by activating latent viral DNA, but more difficult the demonstration that in all cases the viral DNA was responsible. There was no difficulty, when cell-free virus could be shown to cause cancer in the natural or a secondary host, but to demonstrate this could not be attempted in human subjects. Furthermore, because certain tumours such as those of the mammary gland are hormone dependent, it is difficult to provide conditions suitable for its growth in tissue culture.

A further complication has made research difficult. Virus can often be demonstrated in young tumours, but disappears as the tumour grows older. As we have seen, this situation is paralleled, when cells grown in tissue culture are infected with oncogenic viruses. Often, only a minority of cells in the culture are transformed, the majority reacting to viral infection in the normal way by shedding complete virions and being destroyed in the process. The only cells to survive and replicate are then the transformed – or neoplastic – ones, from which complete virus particles can no longer be recovered although viral DNA is present. Viral DNA can be often identified in tumour tissue by its characteristic nucleotide sequences; often, however, it cannot, even when its presence seems assured or it seems that it must have been present. Two possible causes for this are proposed. First, it is believed that analytical techniques may not be sufficiently refined to identify viral DNA when present in very small quantities. Secondly, as seen earlier, the 'hit and run' principle has been suggested, whereby an initial viral intrusion may irreparably damage the DNA control of the cell, which will remain neoplastic even when viral DNA has been eliminated. This argument is at variance with that which asserts that for a cell to become neoplastic new genetic input is required, and would seem to be self-defeating; nevertheless, it is not inconsistent with the idea that only small undetectable amounts of DNA might persist in the cell. In this case, the presence of antibodies in the serum of the host against viral antigens will determine whether there has formerly been or currently is a viral presence.

Before pursuing this theme further, let us consider some of the techniques, many of them elegant and sophisticated, that have been developed to demonstrate a viral presence in neoplastic cells.

Early studies, which showed that malignant tumours were transmissible – as discussed in previous chapters – achieved success by the transplantation of tumour tissue into susceptible hosts. Such experiments do not show whether the cells themselves become transplanted and thus parasitic, or whether the success of the transplant is due to some infectious agent present in them. It is only when cell-free filtrates are shown to cause neoplasia that a viral etiology can confidently be asserted. The introduction of antibiotics enabled the widespread use of tissue cultures to be employed. It was then found, that in a number of cases tumour cells, which were initially non-transmissible, became transmissible after sub-culture; in addition cell-free preparations from such cultures would in time develop infective properties. Even so, an important genetic factor was involved, and receptive strains of animal had to be used if positive results were to be achieved. With chickens numerous breeds were already available; with mice it took twenty years to breed susceptible and resistant strains. In tissue culture work, investigators had one great advantage. Neoplastic cells are immortal, whereas untransformed cells degenerate and die after a few transplants in tissue culture. Neoplastic cell lines may, therefore, be maintained indefinitely in culture, and often eventually produce virus capable of causing neoplasia in susceptible hosts.

The most satisfactory way to demonstrate the presence of virus in tumour tissues is to identify or recover complete viral particles. In young tumours, as stated above, viral particles can sometimes be found by electron microscopy, so that the virus can be identified and classified. If such particles were present in 100% of neoplastic tissues and absent from all normal tissues, this would provide strong presumptive evidence of an etiological relationship, but it could still be argued that the virus was a 'passenger' having found suitable conditions for growth and survival in the transformed tissues; the investigator is still left with the onus of proof that the virus is the actual cause of the trouble. If the same virus transforms cells in tissue culture, the argument for an etiological relationship is strenthened, but does not become absolute, unless it can be shown that the virus induces neoplasia in the living animal. In man this cannot be done, unless the virus induces neoplastic changes in some related host. If the presence of virus or virus DNA can be demonstrated in, say, only 30% of tumour tissues, as with cervical carcinoma of women, evaluation of the results becomes difficult even though the virus can be shown to have oncogenic potential. Evaluation becomes more difficult, when it is found that some, though a lower proportion of normal persons also carry the virus. Particles of the oncorna B group viruses can frequently be found in human breast tumour tissues, though never in healthy mammary tissue; the particles closely resemble those of mouse mammary tumour virus (MMTV). Although absent from healthy mammary tissues, similar particles are often present in milk from healthy mammary glands, and for this reason it is difficult to postulate a causal relationship between the virus and breast cancer. Even so, it seems likely that this virus is the cause of some human breast tumours, though – as seen in an earlier chapter – some may be associated with an ocornavirus of the D group, the Mason–Pfizer Monkey Virus (MPMV).

While infective virus may appear in tissue cultures of tumour cells after some generations, other methods may be used to demonstrate its presence. A neat and successful refinement of this technique is the co-cultivation of the tumour cells with 'permissive' cells of another species; if successful, complete viral particles appear in the alternative cells, in which development is not inhibited. Sometimes the healthy cells are separated from the tumour cells by a semi-permeable membrane; the virus can then pass through the membrane, but the two cultures are kept separate.

In those cases in which attempts to demonstrate mature viral particles are unsuccessful, attempts can be made to extract the DNA of the host cell and demonstrate the presence of viral DNA by analytical methods. As we have seen, such attempts are often also unsuccessful, and other methods must be used to determine whether there is or has been a viral presence. Analytical methods may then be used to discover viral proteins, the so-called 'early antigens', which are transitional or permanent structural features of the viral core. Evidence of viral presence can also be sought by further analyses, as for viral 'messenger' RNA in the cell cytoplasm or for viral enzymes, such as polymerase or transcriptase.

These highly sophisticated techniques have all been used with success in obtaining evidence of viral presence in tumour tissues; where one or more may fail, another may be successful, so that a number of different analyses may need to be done. These analyses may be further reinforced by immunological methods. Serum samples from patients can be tested against suspect viruses, but positive reactions may be weak against endogenous viruses, because of tolerance acquired during foetal life. Positive results against exogenous viruses require to be carefully interpreted. For example, all patients suffering from Burkitt's Lymphoma carry antigens against Epstein–Barr Virus (EBV), but so do a proportion of healthy persons. This is regarded as strong evidence for a causal relationship between EBV and Burkitt's Lymphoma, but shows that some persons become infected without developing symptoms. In addition to tests on the blood serum, tumour tissues may also be examined for presence of antibodies developed against viral proteins or enzymes. For this purpose immuno-fluorescence techniques have an especial value. Immunological techniques may be used to detect antibody developed against entire viral particles, in which case they will reveal sensitivity to coat proteins or enzymes. They can also be used to detect sensitivity to early developmental or core antigens or to polymerases and transcriptases. There is thus an array of tests which may be done, each of which will give different information regarding the presence, past presence and activity of a virus. Immunological techniques also permit the differentiation of closely related strains of virus and their grouping in logical sequence; such is assisted by measurement of viruses under the electron microscope. Such studies can be further enhanced in tissue culture by studying the extent to which related viruses 'recombine' by antigen exchange to form hybrids.

The hypothesis that all neoplasias are utlimately dependent on the presence of viral DNA in the host cell genome can certainly not as yet be discarded, and there is much positive evidence to support it. If correct, then the solution of the problem lies with the cell/virus relationships about which a great deal more information needs to be acquired. It is desirable, too, that more information should be acquired about the relationships of related viruses with each other, as when viruses recombine to form hybrids. Just how, for example, do attenuated chicken viruses and turkey viruses of Marek's Disease prevent the development of neoplasia without inhibiting infection? Do they remove the oncogene from the wild virus? Do they stimulate the activity of humoral or cellular immune systems? Do they stimulate the production of Interferon? Were the answers to these questions available, prophylactic or therapeutic means might be discovered by which oncogenesis could be suppressed in such cancers as are the result of viral gene contamination. A peculiar aspect of the virus cancer problem lies in the random distribution of oncogenic properties both amongst viral groups and within the groups themselves. While the papova- and adenovirus groups are similar, they are far removed from the herpes- and oncornavirus groups, which themselves differ greatly from each other. Two points of similarity can, however, be discerned: 1. they are all viruses, which tend to become persistent symbiotes in the natural host, and 2. the cells which they transform tend to be alien cells, that is cells other than those in which they actively replicate and cause acute infectious conditions. This statement requires to be qualified in the sense that with some oncogenic viruses the cells require to be 'alienated' by co-carcinogenic influences; alternatively, alien cells may only be found following infection of some host other than the natural one. This suggests that cell transformation by viral genes results from an abortive attempt on the part of the virus to live symbiotically in cells to which it is not well adapted to the purpose. This supposition follows from what has been written earlier. The hypothesis that cancers are the result of viral genetic information in the host genome can, therefore, be augmented by postulating that this occurs only when this is present in 'alien' or 'alienated' cells.

Meanwhile, two important aspects of the problem require to be studied. The first of these is the nature of the oncogene; secondly, the relationships and interference phenomena, which occur between two co-infecting related viruses.

The part played by viruses in causing human cancers may still be in doubt though there are good reasons for supposing them to be involved. The studies of recent years have at least achieved one outstanding success, namely the demonstration of the commensal link between viral and vertebral genes on a long term evolutionary basis. The earliest of such associations must date back to the beginning of the Palaeozoic Era, some 550 million years ago, or even earlier. During this vast time, the viruses have evolved *pari passu* with the host, but still retained a separate identity. Although only their genes are transmitted from generation to generation, when conditions are right they can still replicate as complete infectious viral particles. This means that they have retained the power to manipulate the genetic system of the host, forcing it to construct viral DNA and proteins. The random existence of oncogenes in four of the numerous groups of viruses has still to be explained, including the reasons for their presence and the means by which they induce aberrant growth of the host cells.

If the viral genes are, indeed, responsible for the cell mutations and transformations which result in cancer, then a great deal more needs to be known about the ways in which they are inherited and about the relationships of host and viral genes. Are the viral genes 'silent', simply replicating in step with the host genes, or do they play some part in cell differentiation or other activities? Are they concerned with cell mutations and do they in this way, as some virologists have suggested, play some part in evolutionary processes? Are they present equally in both sexes, and are there differences of homozygosity and heterozygosity? If so, what is the significance to the cancer problem? Is it possible that cancers develop, as with other genetic diseases, in those unfortunate individuals who are homozygous, inheriting similar cancer genes from both parents? In what chromosomes are the cancer genes located, or do they part company and become located in different chromosomes?

The monoclonal nature of cancers, if correct, presents a further problem. It is difficult to believe that exposure to radiation or massive doses of carcinogens would affect only one cell in a tissue; this is not what happens in tissue cultures and surely all cells, which are at risk, must be affected equally. Why a tumour should contain cells with either paternal or maternal X chromosomes, but not both, is difficult to explain, although the virologist may claim it to show that viral genes were inherited from only one parent. No means appears to have been discovered to ascertain whether cancers in males are also monoclonal.

When the genes of a virus enter the DNA of the host, two changes occur. A segment of host genome is removed and with it the genetic information it contains; secondly the genetic information of the viral genes is added in substitution. With those viruses, which cause infectious diseases, such as influenza or the common cold, the host cell is manipulated in such a way that new infectious viral particles are constructed; they either infect further healthy cells or are shed in discharges, which can transmit infection to new individuals. In the process, the host cell is destroyed. We do not know what happens, when viruses become commensal. One must suppose that the activities of the replicatory genes of the virus are repressed by the same mechanisms, by which the activities of host genes are repressed. It is not difficult to envisage a situation in which there would be derepression of some viral genes, but not of others; the oncogenes might be derepressed, but not the replicatory, as happens with temperature sensitive mutants. This is a plausible explanation, though it does not answer the question as to why some viruses possess oncogenes and what is their normal purpose? The causation of cancer is plainly an aberrant activity, since the virus is destroyed with the host and their are no survival advantages.

These questions require answers, if a viral involvement in cancer is to be understood and if the viral theory is to be substantiated. There are many other questions that need to be answered, in particular the mechanisms and occasions of gene exchange between related viruses and between host and virus. The progress achieved so far can only excite the admiration of those of us, who study the results, and give us confidence that the remaining questions will be answered in time. (*Vide* Vigier, P., 1974, for a discussion of cryptic oncogenic viruses.)

That some groups of viruses have commensal relationships with vertebrates should not surprise us. Indeed, it would be surprising if it were not so. All other groups of microorganisms play a part in the lives of multicellular creatures. Without commensals, bacteria, fungi and protozoa, the vertebrate animal could not survive. Many biochemical processes, essential to digestion and other metabolic activities, require enzyme systems lost by higher groups of animals. Are viruses, we ask ourselves, unique in that they are wholly pathogens, or do they too perform some useful functions in the body politic? We know now that they are commensally associated with vertebrate cells, but what functions they perform, we do not know. Commensal bacteria and fungi, at times of stress or immunological embarrassment, cause active diseases because the body loses the power to control them. It is probable that viruses, normally harmlessly or helpfully resident, may do the same. It is a good, though speculative, parallel and not necessarily correct.

It is not easy to make coherent sense of the cancer story, when so many eminent experts disagree and there remain so many unanswered questions. On the evidence, which I have attempted to present in these pages, it is undoubtedly a valid hypothesis that all malignant cancers result from the activities of viruses in one way or another. It is profoundly to be hoped that this is so, because in that case there will be better hopes of finding ways to protect people from them and to cure them.

In conclusion, I wish to draw attention to the curious gradation of the means by which oncogenic viruses come to infect their hosts. In the horizontally transmitted cancers, some are transferred as with any other pathogens by direct or indirect contact from an animal carrying the infection to one that is uninfected. Such happens with Marek's Disease of poultry, with the papillomas of rabbits and cattle, with the lymphomas of monkeys resulting from infection with a herpesvirus from a closely related species, and with infectious mononucleosis of man. Then there are the venereally transmitted cancers, of which the venereal tumours of dogs are the outstanding example; however, the virus of mouse mammary carcinoma is also transmitted from male to female in this way, and also the HSV_2 and cytomegaloviruses of man, both suspected of being concerned with neoplasias. The next step in the gradation is the transmission of mouse mammary carcinoma virus in the mother's milk and the neonatal transmission of commensal herpesviruses. The latter are in reality a form of vertical transmission in that a long-term evolutionary association has developed between host and virus. The true vertically transmitted viruses follow next in sequence, where the viral genes become a part of the genetic apparatus of the host. This has happened only with the oncornaviruses, but it is of great interest that a number of these viruses are still horizontally transmitted, suggesting that there has been some process of evolutionary adaptation between virus and host.

Horizontally transmitted viruses would be expected to cause disease in 'clusters' or epidemics, and indeed they do. Since such clusters do not usually occur in human cancers, the implication is that – if viruses are involved – they are those which are acquired shortly after birth or in the parental genome.

Such an adaptive gradation in the means, by which pathogens are spread, is not without parallel. The trypanosomes provide a typical instance. Some are spread mechanically by biting flies, drawing blood from one and then quickly settling on another, into which some blood from the first is injected during the fly's feeding; such is the method with *Trypanosoma evansi,* a parasite of horses and camels. Other trypanosomes are transmitted by biting flies or other vectors after development in the intermediate host; *T. lewisi* of rats is transmitted by fleas by way of the flea's faeces, which are licked off the fur and ingested; with *T. congolense* of cattle, the trypanosomes develop in the salivary glands of tsetse flies; with *T. gambiense* and *T. rhodesiense,* which cause sleeping sickness in man, the trypanosomes develop in the gut of the tsetse fly and then pass to the salivary glands. With *T. equiperdum,* the cause of dourine in horses, the trypanosomes are transmitted venereally. There thus occurs a graded sophistication in the modes of transmission, which is generally regarded as indicating an increasing adaptation in the host/parasite relationship. This is similar to what has been noted with the oncogenic viruses and may indicate a less advanced relationship with those viruses, which are the more recently adapted to their hosts. It indicates in any case an epidemiological pattern, which is not uncommon and which makes the more plausible the belief that cancers are caused by infectious agents like other kinds of disease.

Reference

Vigier, P. (1974). Replication and integration of the genome of oncornaviruses. *In* (Kurstak, E, and Maramorosch, K., eds) 'Viruses, Evolution and Cancer'. New York and London: Academic Press

Appendix

A HYPOTHETICAL MODEL FOR CARCINOGENESIS

The hypothetical model, proposed below, results from reflection on the facts of carcinogenesis as revealed by the comprehensive survey of cancers in a spectrum of the animal kingdom that is wider than normal.

I Causes of cancer

The onset of the cancer process is initiated by the presence of 'alien' or 'alienated' genes in somatic cells. Alien genes result from the insertion of viral genes into the host genome; alienated genes from the presence in the genome of host genes, the functions of which have been altered by faulty DNA copying and repair.

The viral genes are derived from one of the following: 1. exogenous viruses, acquired by horizontal transmission, or 2. endogenous viruses acquired by vertical transmission, or 3. recombinant viruses, resulting from gene transfers between exogenous and endogenous viruses.

The presence of alien or alienated genes in somatic cells causes mutagenesis, but not carcinogenesis which is evidently a sequential process resulting from other factors.

II The effector system

When cancer develops further, cell growth inhibition is impaired because of changes in the receptor systems on the surfaces of the cell membranes (plasmalemma), the chemical composition of which is altered by viral activity (RNA viruses) or faulty maintenance by disordered genes.

III The development sequences of carcinogenesis

Both somatic and viral theories of cancer are incorrect in suggesting that carcinogenesis is caused by altered DNA in somatic cells. That there is an evolutionary sequence in the carcinogenic process results from the study of similar cancers in different hosts, and evidence suggests that the sequential events may be mediated by differing influences. If so, an understanding of them, and the causes of each, might indicate means by which the process could be interrupted or diverted. The sequences of carcinogenesis may be suggested as follows.

1 *The infectious phase (in viral cancers)*

(*a*) *Exogenous viruses* Virus is transmitted horizontally from subjects shedding virus or by experimental procedures. The virus of Marek's Disease of chickens is acquired by inhalation of feather dander from a bird suffering from a disease of typically infectious nature. The Epstein–Barr Virus (EBV), in the causation of infectious mononucleosis, is transmitted from man to man (or woman!) by close oral contact. *Herpesvirus simplex* 2, believed to be the cause of cervical carcinoma, is transmitted venereally. Evidence from the chicken and feline leukaemias and sarcomas indicates that there is an inflammatory phase, during which the body defences are mobilised.

(*b*) *Endogenous viruses* Viral genes are present as inserts in the host genome, where they divide *pari passu* with the host genes, and appear normally to be subject to the same restraints. They may appear as complete replicating particles in the fertilised ovum, the blastocyst, the placenta and the neonate. They are sometimes present in *early* tumour tissues and in the milk of mice and women with malignant breast tumours.

(*c*) *Neonate infections* Neonate infections are acquired from the mother either in the milk or by contact during the interval between birth and weaning. Though they resemble endogenous infections, no immune tolerance has been acquired during foetal life, so that their presence may more easily be demonstrated serologically. Masked endogenous or neonate viruses can sometimes be recovered as replicating particles by co-cultivation of host cells with permissive cells of another species.

2 *The inflammatory phase*

With exogenous viruses, even those with oncogenic potential, the virus enters the target cells. In some hosts, infection takes the usual course of: (*a*) cell destruction and death of the host, or (*b*) elimination of the virus and immunity, or (*c*) cryptic infection. Where cryptic infection occurs, the virus may again be stimulated to proliferate, e.g. varicella/zoster, chicken pox in early life, shingles later. Older people develop antibody to EBV, but do not suffer from IM. In Marek's Disease, neuro-lymphomatosis does not develop in vaccinated chickens, though they acquire infection.

3 *Pre-cancerous stage*

Prior to malignancy, cells undergo mutagenic changes, which may proceed no further. Certain chemicals cause mutagenic changes in cells, but oncogenesis may only occur many years later in response to a further challenge by a different chemical. Sufferers from *Herpes genitalis* may show recognisable morphological changes in the cervical epithelium recognised as pre-cancerous, but cervical carcinoma may not ensue. IM patients recover from clinical symptoms and the B lymphocytes are morphologically normal; they have, however, the ability to proliferate in tissue culture in the lymphoblastoid stage, provided that EBV genes are present in the genome. Some other factor must induce the onset of Burkitt's Lymphoma or naso-pharyngeal carcinoma (NPC).

4 *Benign phase*

Benign tumours are self-limiting and sometimes regress, as with the human wart and papilloma viruses. Malignant tumours sometimes regress spontaneously, as with RSV in adult Old World monkeys. On the other hand, benign tumours may sometimes progress to malignancy, as with the rabbit and bovine papillomas; with the latter, a toxic factor consumed with bracken is blamed as the secondary influence. In young tumours caused by viruses, mature viral particles can often be found, though not in older tumours. This same phenomenon is seen also in tissue cultures, in which there is competition between cells with proliferating virus and those with mutagenic virus. In young tumours, too, inflammatory cells may be an element, but they disappear as the tumour becomes older.

5 *Localised malignancy*

Neoplasms, even the most deadly may remain localised, in which case they are confined to the 'base cells' of the tumour, that is the original cells to have suffered mutagenic changes. If so, they can be excised, and with ray and drug treatments the prognosis is good. A variation of this is the tumour which remains confined to the base cell, but is widely distributed throughout the body. Again some of these, such as lymphatic leukaemia, can be treated with chemical agents with a high degree of success.

6 *Metastasising malignancies*

When metastases occur, cells other than the base cells have usually become involved, and mixed tumours develop. By some means, cells other than those of the original clone participate in the neoplastic process. Treatment is usually inneffectual and the patient dies as a result of widespread malignant tumours in a number of vital organs. Human breast and bowel tumours sometimes remain localised, and treatment is successful; all too often, alien metastasising cells become involved and treatment is hopeless. EBV tumours, BL and NPC, are peculiar in that the former although often becoming generalised, is confined to the base cells. NPC is carcinomatous, so that cells of epithelial origin have becom involved as well as the B lymphocytes. It is significant, in this respect, that BL responds to treatment with cytotoxic drugs. It is vital to an understanding of the carcinomatous process, to discover the influences under which it spreads from the base cells to cells with a different origin.

IV Conclusion

In the proposed hypothesis, it does not matter, whether the 'rogue' genes responsible for the onset of carcinogenesis are of viral or somatic origin. If viral genes are responsible, as seems to be likely in most cases, an additional area of attack is afforded, such as is successful with Marek's Disease vaccines. The importance appears to lie with the sequence of events, each of which may be mediated by a different causative factor. Each needs to be studied independently, if necessary in different animals so as to obtain a composite picture.

Further Reading

The books listed below have been helpful to the author in preparing this work. They are by no means a complete list of all works on cancer etiology

Biggs, P. M., de Thé, G. and Payne, L. M. (1972). 'Oncogenesis and Herpesviruses.' Lyon: International Agency for Research on Cancer

Blough, H. A. and Tiffany, J. M. (1980). 'Cell Membranes and Viral Envelopes', 2 vols. London and New York: Academic Press

Burnet, Sir M. (1978). 'Endurance of Life'. University Press Melbourne

Cairns, J. (1978). 'Cancer, Science and Society'. San Francisco: Freeman

Cochran, A. J. (1978). 'Man, Cancer and Immunity'. London: Academic Press

Gross, L. (1970). 'Oncogenic Viruses', 2nd edn, 991 pp. Oxford: Pergamon Press

Kaplan, A. S. (ed.) (1973). 'The Herpesviruses'. New York and London: Academic Press

Kurstak, E. and Maramorosch, K. (eds). (1974). 'Viruses, Evolution and Cancer'. New York and London: Academic Press

National Cancer Institute (1975–1978). 'The Virus Cancer Program'. Washington: US Dept. of Health, Education and Welfare, Bethesda, Md.

Author Index

Page numbers in italics indicate reference page numbers

Subject Index